Basic Cost Engineering

COST ENGINEERING

A Series of Reference Books and Textbooks

Editor

KENNETH K. HUMPHREYS, Ph.D.

Consulting Engineer
Granite Falls, North Carolina

1. Applied Cost Engineering, *Forrest D. Clark and A. B. Lorenzoni*
2. Basic Cost Engineering, *Kenneth K. Humphreys and Sidney Katell*
3. Applied Cost and Schedule Control, *James A. Bent*
4. Cost Engineering Management Techniques, *James H. Black*
5. Manufacturing Cost Engineering Handbook, *edited by Eric M. Malstrom*
6. Project and Cost Engineers' Handbook: Second Edition, Revised and Expanded, *edited by Kenneth K. Humphreys*
7. How to Keep Product Costs in Line, *Nathan Gutman*
8. Applied Cost Engineering: Second Edition, Revised and Expanded, *Forrest D. Clark and A. B. Lorenzoni*
9. Managing the Engineering and Construction of Small Projects: Practical Techniques for Planning, Estimating, Project Control, and Computer Applications, *Richard E. Westney*
10. Basic Cost Engineering: Second Edition, Revised and Expanded, *Kenneth K. Humphreys and Paul Wellman*
11. Cost Engineering in Printed Circuit Board Manufacturing, *Robert P. Hedden*
12. Construction Cost Engineering Handbook, *Anghel Patrascu*
13. Computerized Project Control, *Fulvio Drigani*
14. Cost Analysis for Capital Investment Decisions, *Hans J. Lang*
15. Computer-Organized Cost Engineering, *Gideon Samid*
16. Engineering Project Management, *Frederick L. Blanchard*
17. Computerized Management of Multiple Small Projects: Planning, Task and Resource Scheduling, Estimating, Design Optimization, and Project Control, *Richard E. Westney*
18. Estimating and Costing for the Metal Manufacturing Industries, *Robert C. Creese, M. Adithan, and B. S. Pabla*

19. Project and Cost Engineers' Handbook: Third Edition, Revised and Expanded, *edited by Kenneth K. Humphreys and Lloyd M. English*
20. Hazardous Waste Cost Control, *edited by Richard A. Selg*
21. Construction Materials Management, *George Stukhart*
22. Planning, Estimating, and Control of Chemical Estimation Projects, *Pablo F. Navarrete*
23. Precision Manufacturing Costing, *E. Ralph Sims, Jr.*
24. Techniques for Capital Expenditure Analysis, *Henry C. Thorne and Julian A. Piekarski*
25. Basic Cost Engineering: Third Edition, Revised and Expanded, *Kenneth K. Humphreys and Paul Wellman*

Additional Volumes in Preparation

Basic Cost Engineering

Third Edition, Revised and Expanded

Kenneth K. Humphreys
Consulting Engineer
Granite Falls, North Carolina

Paul Wellman
Consultant
Ashland, Kentucky

Marcel Dekker, Inc. New York • Basel • Hong Kong

Library of Congress Cataloging-in-Publication Data

Humphreys, Kenneth King.
Basic cost engineering / Kenneth K. Humphreys, Paul Wellman. - 3rd ed., rev. and expanded.
p. cm. - (Cost engineering ; 25)
Includes bibliographical references and index.
ISBN 0-8247-9670-5 (alk. paper)
1. Engineering economy. 2. Costs, Industrial. I. Wellman, Paul. II. Title. III. Series: Cost engineering (Marcel Dekker, Inc.) ; 25.
TA177.4.H85 1996
658.15'5-dc20 95-34362
CIP

The publisher offers discounts on this book when ordered in bulk quantities. For more information, write to Special Sales/Professional Marketing at the address below.

This book is printed on acid-free paper.

Marcel Dekker, Inc.
270 Madison Avenue, New York, New York 10016

Current printing (last digit):
10 9 8 7 6 5 4 3 2 1

PRINTED IN THE UNITED STATES OF AMERICA

To the late Dr. James H. Black,
a dedicated colleague, mentor, and friend

Preface to the Third Edition

All cost engineering dissertations should be updated periodically to reflect the current state of the art. *Basic Cost Engineering* was first written for use as a textbook on capital and operating cost estimation. It was, and still is, intended to introduce cost engineering principles to those individuals who are entering the field and as a reference work and refresher for the experienced cost engineering professional.

Virtually all the material that appeared in the first two editions of *Basic Cost Engineering* is as valid today as it was when it was first published. With the exception of changes necessitated by U.S. tax laws and depreciation regulations, very little material from the second edition required revision. However, since the second edition was published, there has been a decided shift in the needs of industry with respect to cost estimating requirements. Industrial emphasis has moved in many cases from construction of new chemical processing plants to the retrofitting and expansion of existing plants. For this reason, a chapter has been added to discuss estimating of retrofit and expansion projects in existing, operating plants. The portions of the book dealing with factored estimating have also been considerably expanded to make them more comprehensive.

With the steady trend toward economic globalization and international economic development, interest has surged in estimating projects in various locations around the world. While estimating principles as discussed in this book do not vary with location, cost information, labor productivity, construction conditions, taxation, etc., do vary drastically from nation to nation. A chapter has therefore been added to provide some reference sources for international cost data.

Sections of the book on profitability analysis, risk analysis, and sensitivity analysis have also been completely revised and expanded. The discussion now covers the vital subject of life cycle cost analysis, a topic of major concern, particularly in government funded projects.

A new bibliography has been included that cites reference works, journals, and conference proceedings expected to be readily available to the reader for many years to come rather than specific articles from past periodical literature that rapidly become difficult to locate.

Throughout the book numerous problems and questions have been added for the benefit of students and educators. A solutions manual is available for use by college and university professors.

Finally, lest critics begin to complain that the book ignores computers, it doesn't. Computers are vital tools in cost engineering work, but this book is not a guide to computer applications. Rather, it is a basic text to explain and teach the fundamental aspects of cost estimating, not to explain computer applications. One cannot properly utilize modern cost engineering tools such as computers without understanding the underlying principles that apply to cost estimating and control. Robot-like use of estimating programs without being able to discern the reasonableness of the results and the validity of assumptions made in developing those programs is foolhardy. The estimator must first understand estimating principles and procedures before using computerized techniques. That understanding is what this book is designed to teach.

The authors wish to acknowledge the assistance of William R. Barry, CCE, and Dr. Harold E. Marshall, who provided considerable input in the preparation of the third edition. The authors also wish to acknowledge the contributions of Julian A. Piekarski and the late Sidney Katell to the earlier editions of this book. Many of their contributions have been carried forward to this edition.

Kenneth K. Humphreys
Paul Wellman

Preface to the Second Edition

The first edition of *Basic Cost Engineering* was written to fill a void in the literature on cost engineering by providing a basic text covering the principles of capital and operating cost estimating and profitability analysis that assumed no prior knowledge of these topics by the reader.

The first edition succeeded in meeting its objectives but did have a number of deficiencies, which the present edition is designed to correct.

Major changes include considerable expansion of the discussions of capital cost estimating techniques and of profitability analysis and the inclusion of material on risk analysis. The section on operating costs has been revised to include current depreciation guidelines and tax considerations as of January 1986. The appendix on cost engineering literature has been considerably expanded and reformatted to simplify its use in locating literature via keywords.

The bibliography is completely new and with a very few exceptions covers literature published since 1982. This is an exhaustive bibliography of current literature in the cost engineering field and is believed to be the most complete bibliography of its type. Over 600 complete bibliographic citations are provided with an alphabetized keyword cross-list.

Finally, an appendix of cost engineering problems, including a complete cost estimate, has been included for the use of instructors and students. For ease of use, all problems in the book have been placed in this one appendix rather than at the end of the various chapters as was the case with the first edition of *Basic Cost Engineering*.

Other changes include deletion of the former chapter on accounting considerations, which was felt to be peripheral to the main theme of the book, and the deletion of the former appendix of equipment cost charts. The cost charts were deleted because they become obsolete very rapidly and can also be easily misinterpreted. For information of the type included in these former charts, the reader should refer to any of the many excellent annual cost data books that are on the market.

The authors wish to acknowledge the assistance of Julian A. Piekarski in the preparation of the material on profitability and risk analysis and the many contributions of Sidney Katell, who coauthored the first edition.

Kenneth K. Humphreys
Paul Wellman

Preface to the First Edition

Cost engineering is defined as "the application of scientific and engineering principles and techniques to problems of cost estimation, cost control, business planning and management science." Implicit in this definition are problems of profitability analysis, project management, and planning and scheduling of major engineering projects. Cost control, project management, planning, and scheduling are major topics of the first book in this series, *Applied Cost Engineering* by Clark and Lorenzoni. The present book deals with the remaining aspects of cost engineering, cost estimation, and profitability analysis with emphasis on both capital and operating costs. Emphasis is placed upon the evaluation and planning phases of major projects only, and no effort is made in the present work to define techniques for developing detailed analyses of costs of major projects.

Most of the literature in the cost engineering field is rather advanced and assumes a substantial cost engineering background on the part of the reader. The basic principles of cost engineering, those that bridge the gap from classical "engineering economics" to detailed cost estimation and control on major projects, are often brushed over lightly. In truth they are the most critical cost engineering topics which must be understood and used by virtually every engineer and engineering student. These basic principles include preliminary capital cost estimating, operating cost estimating, and profitability analysis. These three topics are the major focus of this book.

Most cost engineering decisions which are made in industry are made on projects for which little detailed information is available; engineering designs have not even begun in most cases, and at best a detailed flowsheet of the proposed process is available. Obviously, with this lack of information, a detailed cost estimate cannot be performed. However, using the basic principles of cost engineering, a preliminary capital and operating cost estimate and a profitability analysis can be made with sufficient accuracy to enable management to decide whether or not to proceed with further development of the project.

The authors wish to acknowledge the valuable assistance of the staff of the Coal Research Bureau, College of Mineral and Energy Resources, West Virginia University in typing, editing, and the preparation of graphs and illustrations for this text.

Kenneth K. Humphreys
Sidney Katell

Contents

Preface to the Third Edition *v*
Preface to the Second Edition *vii*
Preface to the First Edition *ix*

Part I: Capital Cost Estimates

1. Capital Cost Estimating: An Overview 3

   Questions

2. Classification of Capital Cost Estimates 7

   Order-of-Magnitude (Ratio) Estimates / Study (Factored) Estimates / Preliminary (Budget Authorization) Estimates / Definitive (Project Control) Estimates / Detailed (Final) Estimates / Problems and Questions

3. Organizing the Capital Estimate 47

Questions

4. Conceptual Estimating Using MasterFormat 61

Introduction / Work Breakdown Structures (WBS) / Standard Code of Accounts / Rules of Thumb / Conclusion / Nomenclature / Questions

5. Retrofitting and Expansion of Existing Plants 95

Introduction / Accessibility / Plant Safety Considerations / Electrical Power Source and Motor Control Centers / Outdated As-Built Drawings / Pipe-Rack Loading / Control Systems / Availability of Plant Air, Steam, and Water / Civil Engineering Scope / Piping and Mechanical Scope / Other Considerations / Questions

6. Labor Estimates 101

Estimates of Labor for an Order-of-Magnitude Estimate / Estimates of Labor for a Factored Estimate / Estimate of Labor for a Preliminary (Budget Authorization) Estimate / Definitive Estimate Workhours / Detailed Estimate Workhours / Work Breakdown Structure and Labor Planning / Questions

7. Importance of the Capital Estimate 109

Scheduling / Contingency / Questions

8. Sources of International Cost Data 115

Introduction / Limitations of Published Data / Background Sources / International Location Factors / Country Cost Indexes / Cost Data Sources / Summary / Problems and Questions / References and Information Sources

Part II: Operating and Manufacturing Cost Estimates

9. Operating and Manufacturing Cost Estimating: An Overview 141

Questions

10. Estimating Operating and Manufacturing Costs 149

Definitions / Types of Operating Cost Estimates and Estimating Forms / Cost of Operations at Less than Full Capacity / Raw Materials Costs / By-Product Credits and Debits / Utility Costs / Labor Costs / Supervision and Maintenance Costs / Operating Supplies and Overhead Costs / Royalties and Rentals / Contingencies / General Works Expense / Depreciation / Accelerated Cost Recovery System / Modified Accelerated Cost Recovery System / Amortization, Depletion, Insurance, and Real Estate Taxes / Distribution Costs / Problems and Questions / References

11. Transition: Construction to Operation 207

Part III: Financial Analysis and Reporting

12. Profitability Analysis: An Overview 213

Cash Flow / Basic Concepts of Profitability Analysis / Assumptions and Definitions / Methods of Measuring Profitability / Revenue Requirements / Problems and Questions

13. Life-Cycle Costing and Sensitivity Analysis 229

Introduction / How to Measure and Apply LCC / Other Methods and When to Use Them / Uncertainty and Sensitivity Analysis / Problems / References

14. The Finished Report 245

14.1 Final Problem—A Complete Cost Estimate

Appendixes

A. Cost Engineering Terminology 253

B. Design and Costing Specifications for Equipment 277

C. Discrete Compound Interest Tables 295

D. Bibliography 319

Cost Data Books / Periodical Literature / Reference Books

E. Answers to Selected Problems 331

Index *333*

Basic Cost Engineering

Part I

Capital Cost Estimates

1

Capital Cost Estimating
An Overview

Capital cost estimating has been a much maligned vocation due to serious cost overruns on too many construction projects. There are many reasons for the problem, but the estimator must accept at least part of the responsibility. However, there are sufficient causes for this phenomenon to allow some of the blame to be lavishly distributed throughout our plant construction organizations. In fact, these organizations may well be a part of the problem.

Estimating equipment costs does not afford the magnitude of error that installation cost estimates seem to contribute to overall estimating inaccuracies.

Process equipment estimating tends to be dependent upon the "scope of work," process flowsheets, material and energy balances, and a complete descriptive equipment list. Obviously, to have all of these items, considerable preliminary engineering must be completed and properly incorporated into the support for the estimate. With the above information the equipment estimates are normally very accurate.

Difficulties begin to develop in the determination of the bulk accounts such as concrete, piping, instruments, electrical, and so on. In preliminary estimates, these are poorly defined and the engineer must use factors that do not normally include such items as interconnecting piping, electrical distribution, fire water loops, cooling towers, and cooling water distribution. Even the most experienced estimator will occasionally grossly underestimate these accounts.

The other accounts which are often underestimated are sometimes termed the "offsites." These are systems required for plant operations but are not actually processing units. They include steam generation and distribution, water supply, wastewater treatment, environmental systems, administrative buildings, parking lots, fences, raw material receiving, product shipment, etc. In the normal evolution of a project, these systems are engineered last; therein lies part of the problem.

A review of available literature dealing with capital cost estimation indicates a lack of a comprehensive reference on the subject. Therefore, an attempt to fill this need without getting too involved in the numerous related subject areas is the aim of this book.

Capital estimating is as old as civilization and the reasons for estimating have remained much the same from the beginning; that is, before embarking upon a significant construction project or other endeavors requiring expenditures of large sums of money, those concerned need to know and, therefore, ask how much it will cost. One of the rudest awakenings a young science or engineering major receives upon leaving school is that the boss, after hearing how great his or her new idea is, wants to know the costs. In fact, large corporations make capital expenditure estimates for ten years or more in advance to aid in planning corporate strategy.

Cost overruns are not new although such occurrences seem to be more prevalent in recent years. Saint Luke's gospel refers to a man who started to build a house and ran short of funds as a fool. When the United States was very young, the actual cost to build "Old Ironsides" was three times the estimate. Poor estimating probably has a history as long as estimating.

After reading the preceding paragraph, one might simply ask, "If estimating is this bad, why continue to estimate?" Like any other professional product, the poor estimates get the publicity; the estimates that were correct do not. For every extremely poor estimate, probably ten were done well or at least adequately for the purpose intended.

Estimates are prepared and used for different purposes, including the following:

1. Feasibility studies
2. Selection from among alternate designs

3. Selection from among alternate sites
4. Selection from among alternate investments
5. Presentation of bids
6. Appropriation of funds

Estimates may be and are made at all stages of the development of a project from research to production. There are available today good systems and procedures for each stage which, when properly implemented, produce a capital estimate which is acceptable and most useful to management for decision making. However, the estimator retains, throughout all levels of estimating, a relative importance because it is impossible to rid any estimate of some judgments on the part of the person doing the work.

This book reflects the chemical engineering background of the authors, but it is not intended to exclude estimates for other disciplines which may be accomplished using similar procedures.

QUESTIONS

1.1 What part of a capital cost estimate is generally accurate?

1.2 What two areas of capital estimating contribute most to the inaccuracies of an estimate?

1.3 What is the stated purpose of this book?

1.4 What are capital estimates used for?

2

Classification of Capital Cost Estimates

AACE International (formerly the American Association of Cost Engineers) has proposed three classifications of cost estimates. In increasing order of accuracy, the various major types of estimates proposed by AACE are:

Type	Accuracy
Order of magnitude	–30% to +50%
Budget	–15% to +30%
Definitive	–5% to +15%

Other authors have used different labels for the various types of estimates. As a frame of reference and again in increasing order of accuracy, one finds "quickie," "original," "study," "preliminary," "official" or "budget," and "final" or "definitive." Appendix A, a glossary of cost engineering terminology, defines some of these estimate types in more detail.*

* Appendix A covers a great many cost engineering terms in some detail, many of which are not otherwise discussed in this book. All of the terms, however, are commonly encountered in cost engineering work, and thus are included for ready reference.

AACE previously had proposed five classifications of estimates as follows:

1. Order-of-magnitude (ratio) estimate
2. Study (factored) estimate
3. Preliminary (budget authorization) estimate
4. Definitive (project control) estimate
5. Detailed (final) estimate

The two AACE lists illustrate the confusion in terminology relating to classification of estimates, and while they appear to be inconsistent, they really are not if one considers that budget estimates as currently defined by AACE include the previous categories of factored and budget authorization estimates. Similarly, the current category of definitive estimates includes the prior categories of project control and final estimates. Thus the earlier grouping is really an expansion of the current terms. The expanded list of categories is used in the following discussion.

It should be kept in mind that rarely does an estimate fully meet all of the requirements of a particular classification until one gets to the final detailed estimate. That is, estimates normally require combinations of methods for attaining any desired level of accuracy.

One of the risks taken in establishing any capital number is that the estimate produced is usually stripped of qualifying statements and taken as the absolute number. Therefore, capital estimators must be able to communicate effectively the limitations placed on any capital estimate made early in the development of a project.

ORDER-OF-MAGNITUDE (RATIO) ESTIMATES

The order-of-magnitude or ratio estimate costs the least to prepare and has the least accuracy when finished. At best the accuracy of such an estimate is –30 to +50% for an uncomplicated plant and probably worse for a large complex installation. The accuracy is absolutely dependent upon historical data for support of the exponent used in the following equation:

$$C_x = C_k \left(\frac{E_x}{E_k} \right)^n$$

where

C_x = Cost of plant and/or equipment item of size E_x

C_k = Known cost of plant and/or equipment item of size E_k

n = Cost capacity exponent

This cost-capacity equation is often called the six-tenths rule or the seven-tenths rule because the exponent, n, has an approximate value of 0.6–0.7 for many types of plants and equipment. The 0.6–0.7 factor derives from the fact that costs tend to be proportional to surface area of a container while capacity is proportional to volume. In actual practice, the exponential factor varies widely, particularly when applied to total processing systems (plants) as opposed to equipment items. The range of variance is from about 0.3 to greater than 1 under certain unusual circumstances as illustrated by Tables 2.1 and 2.2. Therefore, without extensive historical data, considerable error may be and often is introduced by the choice of exponent.

The use of the cost-capacity formula is also limited in accuracy because the exponent generally is not constant and does change as equipment size increases. For this reason and because of structural differences as equipment is increased in size, the formula should generally be limited for maximum accuracy to capacity ratios of 2:1 or less and, in any event, should never be used at ratios above 5:1.

In using the cost-capacity equation, the data must first be adjusted to take into account the time that the data were originally obtained because the base data are generally historical. Cost indexes are available to bring such data up to date. The equation for bringing cost data up to date is:

$$C_A = C_B \left(\frac{I_A}{I_B} \right)$$

where

C_A = Cost of plant and/or equipment at index value I_A

C_B = Cost of plant and/or equipment at index value I_B

Table 2.1 Cost Capacity Exponents for Scaling Costs of Plants

Plant or process unit	Exponent
Refrigeration, centrifugal	0.68
Refrigeration (incl. auxiliaries)	0.85-0.96
Refrigeration (no auxiliaries)	0.81
Desalting crude oil	0.60
LP-gas recovery in refineries	0.70
Solvent extraction units	0.73
Polymerization, small plants	0.73
Polymerization, large plants	0.91
Solvent dewaxing units	0.82
Lubricating oil manufacture	0.89
Clay treating, percolation (incl. regeneration)	0.55
Clay treating, contact	0.53
Cracking, catalytic	0.83
Cracking, thermal (also reforming)	0.51-0.70
Cracking, thermal, 2-coil	0.79
Natural gasoline plants	0.73
Gas-cycling plants	0.69
Desulfurization, cata. (old)	0.80
Hydrodesulfurization (recent)	0.57
Hydrodesulfurization of gases	0.75
Desulfurization of gases	0.41
Coking (old data)	0.72
Coking (recent)	0.81
Topping (old data)	0.58-0.67
Topping (recent)	0.62
Topping	0.64
Vacuum distillation (old data)	0.80
Vacuum distillation (recent)	0.57
Vacuum flash (old data)	0.64
Vacuum flash, small units	0.41
Vacuum flash, large units	0.32
Hypersorption	0.43
Coke oven gas separation	0.82
Steam generation, package units	0.61
Steam generation, refinery (no auxiliaries)	0.72
Steam generation, large, 200-psi	0.61
Steam generation, large, 1000-psi	0.81
Steam generation, indoor	0.66

Table 2.1 *(Continued)*

Plant or process unit	Exponent
Steam generation, outdoor	0.61
Power generation, 2,000-20,000 kW-hr	0.88
Power generation, oil field, 20-200 kW-hr	0.50
Cooling towers	0.64
Pipelines (cost vs. dia. squared)	0.72
Gas dehydration (field practice)	0.61
Water-gas manufacture	0.81
Fischer-Tropsch (complete)	0.77
CO and CO_2 removal from hydrogen	0.74
Peterson sulfuric acid towers	0.73
Soybean extraction plant	0.70
Water-gas shift conversion	0.69
Sulfuric acid (contact process)	0.66
Sulfuric acid, general	0.78
Sulfuric acid, akylation acid	0.89
Sulfur from H_2S	0.64
Sulfur from H_2S, package plants	0.40
Nitric acid or urea	0.93
Oxygen plants	0.65
Styrene plant	0.65
GR-S synthetic rubber	0.63
Ammonia, alone	0.90
Ammonia, and nitric acid or urea	0.98
Ammonia, and other products	0.95
Ammonia, most complete	1.02
Ethylene	0.86
Hydrogenation stalls, vapor or liq.	0.32
Chlorine plants, electrolytic	0.75
Reforming, cata.	0.62
Aklylation, small plants	0.67
Alkylation, large plants	0.60
Refineries, small	0.57
Refineries, large	0.67
Hydrogen mfr., steam hydro	0.72
Hydrogen sulfide removal	0.55
Average:	0.69

Table 2.2 Cost Capacity Exponents for Scaling Costs of Process Equipment

A. According to Peters and Timmerhaus[a]

Equipment	Size	Exponent
Blender, double cone, rotary c.s.	50–250 cu. ft.	0.49
Blower, centrifugal	1,000–10,000 cfm	0.59
Centrifuge, solid bowl, c.s.	10–100 hp drive	0.67
Crystallizer, vacuum batch, c.s.	500–7,000 cu. ft.	0.37
Compressor, reciprocating, air-cooled, two-stage, 150 psi discharge	10–400 cfm	0.69
Compressor, rotary, single-stage, sliding vane, 150 psi discharge	100–1,000 cfm	0.79
Dryer drum, single vacuum	10–100 sq. ft.	0.76
Dryer drum, single atmospheric	10–100 sq. ft.	0.40
Evaporator (installed) horizontal tank	100–10,000 sq. ft.	0.54
Fan, centrifugal	1,000–10,000 cfm	0.44
Fan, centrifugal	20,000–70,000 cfm	1.17
Heat exchanger, shell-and-tube, floating head, c.s	100–400 sq. ft.	0.60
Heat exchanger, shell-and-tube, fixed sheet, c.s	100–400 sq. ft.	0.44
Kettle, cast iron, jacketed	250–800 gal	0.27
Kettle, glass-lined, jacketed	200–800 gal	0.31
Motor, squirrel cage, induction, 440v, explosion proof	5–20 hp	0.69
Motor, squirrel cage, induction, 440v, explosion proof	20–200 hp	0.99
Pump, reciprocating, horizontal, c.i. includes motor	2–100 gpm	0.34
Pump, centrifugal, horizontal, cast steel, includes motor	10,000–100,000 gpm × psi	0.33
Reactor, glass-lined, jacketed (without drive)	50–600 gal	0.54

Table 2.2 (*Continued*)

Equipment	Size	Exponent
Reactor, s.s. 300 psi	100–1,000 gal	0.56
Separator, centrifugal, c.s.	50–250 cu. ft.	0.49
Tank, flathead, c.s.	100–10,000 gal	0.57
Tank, c.s., glass-lined	100–1,000 gal	0.49
Tower, c.s.	1,000–2,000,000 lbs	0.62
Tray, bubble-cap, c.s.	3–10 ft. diameter	1.20
Tray, sieve, c.s	3–10 ft. diameter	0.86
Average		0.60

B. According to the U.S. Bureau of Mines[b]

Equipment	Exponent
Heat exchangers	
0–3,000 sq. ft.	0.62
3,000–4,000 sq. ft.	0.95
500–5,000 sq. ft.	0.70
U-tube, 0–1,000 sq. ft.	0.53
U-tube, 1,000–4,000 sq. ft.	0.98
Air aftercoolers	0.62
Refrigeration coolers	0.68
General	0.51–0.59
Condensers	
Copper	0.54
Stainless	0.51
Tanks	
Oil-field	0.57
Rundown	0.60
Terminal cone-roof	0.70

Table 2.2 (*Continued*)

Equipment	Exponent
Floating-roof	0.73
Spherical, 15 psi	0.63
Spherical, 100 psi	0.56
Cylindrical pressure	0.53
API flat-bottom not erected	0.63
Redwood, 2–in.	0.59
Redwood, 3–in.	0.63
Horiz. cyl., 250–2,000 gal.	0.50
Horiz. cyl., 2,000–12,000 gal.	0.87
Carbon steel	0.44
Stainless	0.68
Spherical carbon steel	0.70
General	0.48
Average	0.61
Filters	
Sweetland	0.58–0.67
Sweetland, 2–in. spacing	0.63
Sweetland, 4–in. spacing	0.60
Oliver	0.69
Oliver, steel or wood	0.67
Vacuum	0.48
Plate and frame, c.i.	0.57–0.68
Plate and frame, wood	0.35
Plate and frame, recessed	0.58
Average (except wood)	0.62

Table 2.2 (*Continued*)

Equipment	Exponent
Towers	
Carbon, const. diameter	0.70
Carbon, const. height	1.00
Stainless, plate	0.63
Stainless, packed	0.55
Compressors, reciprocating	
Single-stage, 20–30 psi	0.67
Two-stage, 100 psi	0.75
One- and two-stage	0.90
Three- and four-stage	0.64
Gas-engine drive	0.89
General	0.87
Compressors	
Turbo 3,500 rpm	0.50
Centrifugal	0.78
Average	0.75
Electric motors	
440v	0.84
2,200v	0.38
Squirrel, dripproof	0.84
Wound rotor	0.66
1–10 hp	0.32
10–100 hp	0.55–0.60
Average	0.61
Pumps:	
Centrifugal	
Over 250°F	0.42
Under 250°F	0.46

Table 2.2 (*Continued*)

Equipment	Exponent
Horizontal, c.i.	0.67
Vertical, c.i.	0.98
Horizontal, stainless	0.70
Other pumps	
Plunger or triplex	0.63
Duplex	0.59
Simplex	0.34
Positive-displacement	0.70
Plunger	0.75
Piston	0.69
Triplex	0.60
Average	0.61
Gas holders	
No foundations	0.75
General	0.51–0.58
Average	0.61
Burners	
Oil	0.82
Gas	1.06
Belt conveyers	
12–in. width	0.80
18–in. width	0.85
24–in. width	0.90
Screw conveyors	
4–in. diameter	0.90
12–in. diameter	0.83
20–in. diameter	0.73
Average	0.83

Table 2.2 (*Continued*)

Equipment	Exponent
Evaporators	
Std. vertical or horizontal	0.53
Forced circulation, c.i.	0.41
Evaporators, long-tube	
Vertical	0.81
Copper surface, c.i.	0.49
Rubber-lined and lead	0.55
Crushers and grinders	
Swing hammer bar	0.75
Jaw and rotary crusher	0.57
Ball mill	0.60
Roller crusher and mikro	0.66
Gyratory	0.70
Rotary cutters	0.62
Average	0.65
Dryers	
Drum	0.41
Rotary, hot-air	0.87
Rotary, flue-gas	0.92
Spray	0.51
Furnaces	
Topping stills	0.72
Steel tubes	0.48
Chrome steel tubes	0.80
General	0.70
Cracking or pressure	0.61
Dowtherm	0.36
Average (except Dow)	0.66

Table 2.2 *(Continued)*

Equipment	Exponent
Gas producers	0.43
Structural steel	
Tower platform	0.74
Tubestill structure	0.68
Average	0.71
Waste-heat boilers	0.90
Average of all	0.65

Note: When selecting exponents for similar types of equipment appearing in both of the lists above or listings in Chapter 4, choose the equipment description which most closely describes the item being estimated.

[a] *Source*: M.S. Peters and K.O. Timmerhaus, *Plant Design and Economics for Chemical Engineers*, 3rd ed., McGraw-Hill, New York, 1980, p. 167. Reprinted with permission of McGraw-Hill, Inc.

[b] *Source*: S. Katell, *Cost Engineering in the Minerals Industry Study Guide*, Process Evaluation Group, U.S. Bureau of Mines, Morgantown, WV, undated.

Numerous cost indexes can be used, and some of the major ones are:

1. *Engineering News-Record (ENR) Indexes* for building and construction (1913 and 1967 base years).
2. *Marshall and Swift (M&S) Index* of installed equipment (1926 base year). This index was previously identified as the *Marshall and Stevens Index.*
3. *Nelson-Farrar Refinery Construction Index* (1946 base year).
4. *Chemical Engineering (CE) Plant Cost Index* (base 1957–1959).

Care must be taken to use the proper index to fit the building, structure, or equipment involved in the specific cost analysis. Preferably, a special purpose index (such as the Nelson-Farrar index) should be used if applicable to the project being considered.

Some of the variations in indexes are as follows:

The ENR Indexes are based on costs of structural steel, portland cement, lumber, common labor (the construction index), and skilled labor (the building index).

The *Marshall and Swift Index* covers 47 industries and also contains a composite index.

The *Nelson-Farrar Refinery Construction Index* is based on 40% material and 60% labor in the petroleum and refinery industry.

The *Chemical Engineering Index* is based on equipment, machinery, and supports: erection and installation labor; buildings, material, and labor; and engineering and supervision manpower.

Values of these and other widely used indexes are published regularly in *Chemical Engineering, Engineering News-Record,* and other widely circulated trade magazines.

The use of cost indexes has several limitations, and caution is advised in their use. To cite these potential limitations, one must bear in mind the following:

1. Cost indexes may be limited in accuracy. Two indexes could produce two different answers.
2. Cost indexes are based on average values. Specific cases may be different from the average.
3. At best, accuracy may be limited by applying indexes over a 4- to 5-year period.
4. For time intervals over 5 years indexes are highly inaccurate and should be used for order-of-magnitude estimates only. In any event, use of indexes should be avoided for time spans in excess of 10 years.

It is conceivable that costs of similar equipment or plants are on file, but differ in size from the one under study. With limited cost data, a cost-capacity factor can be applied to determine the approximate cost of the new piece of equipment.

Although there are modifications of the general approach to performing ratio estimates, most of the so-called “back-of-the-envelope-type estimates” use the above methodology in one form or other. It is the primary procedure for preparing an order-of-magnitude estimate.

The advantage of such an approach is that a number can be determined quickly and inexpensively. Such estimates may be used to screen capital project proposals but they are an insufficient base for major economic decisions. Neither should such an estimate be used as a budget estimate or as the first serious effort to get a capital number for a project against which cost overruns are measured later.

STUDY (FACTORED) ESTIMATES

The study or factored estimate, as the name implies, is based upon the use of cost factors. That is, the total installation costs are factored from an estimate of the cost of the "bare equipment" using distributive factors. Rarely is an estimate a purely factored estimate, but it is conceivable that such an estimate could be and is sometimes made.

A factored estimate generally requires process flowsheets (preliminary), materials balances, energy balances, and preliminary determination of plant complexity as applicable to the particular project. For example, process flowsheets are not needed for mine development, but machinery utilization and time cycles would be required. Therefore, in estimating, as in other endeavors, common sense prevails.

The first step in preparing an estimate is to gather all available processing or operating data available from laboratory or pilot plant studies and to utilize this data in preparing a basic flowsheet. The following information should be incorporated:

1. Raw material feed and composition into each operating unit
2. Temperature, pressure at inlet
3. Residence or reaction time in unit(s)
4. Temperature and pressure within unit(s)
5. Composition of intermediate and outlet streams
6. Temperature and pressure of intermediate and outlet streams

A typical flowsheet is shown in Figure 2.1. The information shown on the flowsheet serves as a base in sizing of the major equipment items. As an example, Table 2.3 illustrates the design characteristics of a purification section in a synthetic fuel process. Appendix B provides a listing of design and costing specifications for many types of process equipment.

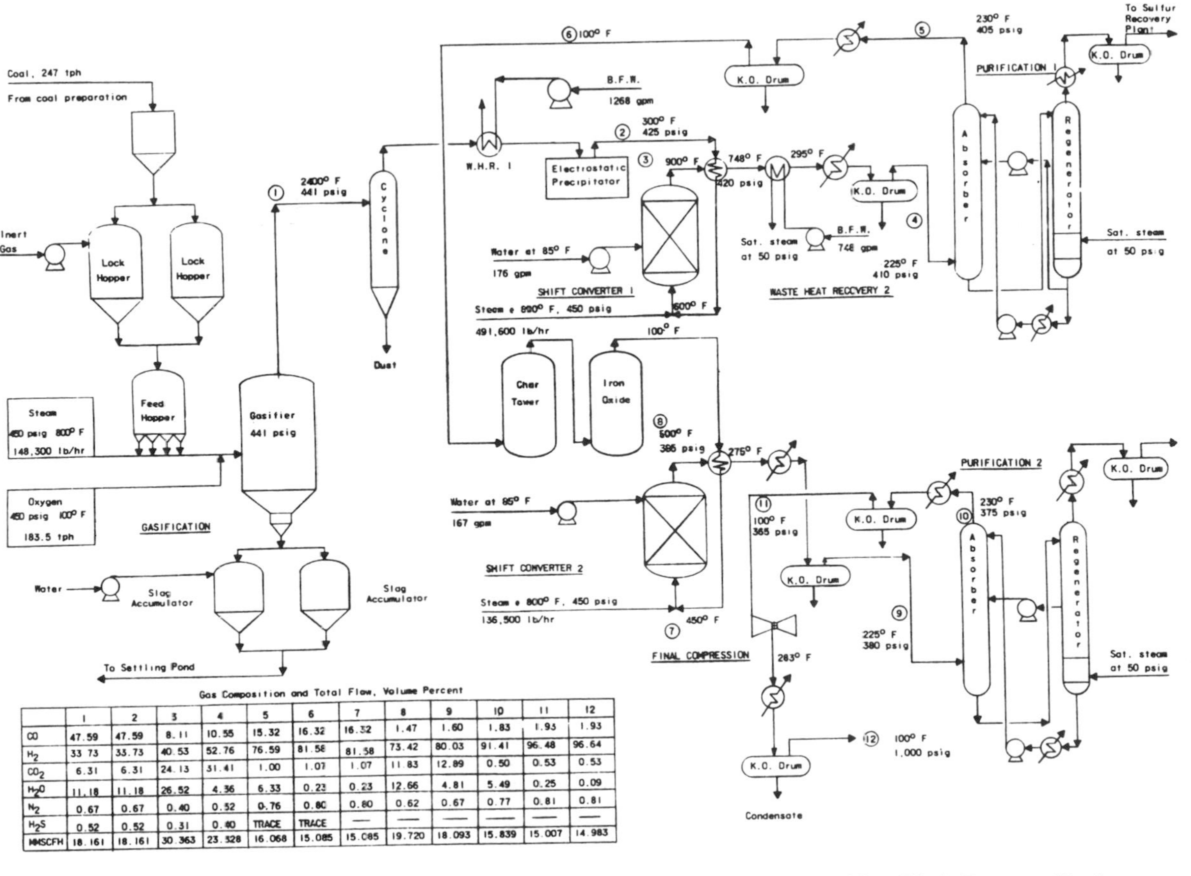

Gas Composition and Total Flow, Volume Percent

	1	2	3	4	5	6	7	8	9	10	11	12
CO	47.59	47.59	8.11	10.55	15.32	16.32	16.32	1.47	1.60	1.83	1.93	1.93
H_2	33.73	33.73	40.53	52.76	76.59	81.58	81.58	73.42	80.03	91.41	96.48	96.64
CO_2	6.31	6.31	24.13	31.41	1.00	1.07	1.07	11.83	12.89	0.50	0.53	0.53
H_2O	11.18	11.18	26.52	4.36	6.33	0.23	0.23	12.66	4.81	5.49	0.25	0.09
N_2	0.67	0.67	0.40	0.52	0.76	0.80	0.80	0.62	0.67	0.77	0.81	0.81
H_2S	0.52	0.52	0.31	0.40	TRACE	TRACE	—	—	—	—	—	—
MMSCFH	18.161	18.161	30.363	23.528	16.068	15.085	15.085	19.720	18.093	15.839	15.007	14.983

Figure 2.1 Process flowsheet for hydrogen production. (*Source*: Report No. 76-6, Process Evaluation Group-MMRD, U.S. Department of the Interior, Bureau of Mines, WV, August, 1975.)

Table 2.3 Detailed Equipment List. Acid Gas Removal

Amine contactor (4 required)
Size: Top, 9′ ID × 29′6″ high; bottom, 12′ ID × 35′6″ high
Operating pressure: 200 psig
Operating temperature: 150°F

Amine regenerator (2 required)
Size: 19′ ID × 84′ high
Operating pressure: 50 psig
Operating temperature: 260°F

Caustic precontactor (2 required)
Size: 2′ ID × 24′ high
Operating pressure: 180 psig
Operating temperature: 120°F

Caustic contactor (2 required)
Size: 4′6″ ID × 61′ high
Operating pressure: 180 psig
Operating temperature: 120°F

Amine knockout drum (2 required)
Size: 12′ ID × 16′6″ high
Operating pressure: 180 psig
Operating temperature: 120°F

Amine flash drum (2 required)
Size: 10′ ID × 30′ high
Operating pressure: 60 psig
Operating temperature: 150°F

Regenerator reflux drum (2 required)
Size: 9′ ID × 11′ high
Operating pressure: 50 psig
Operating temperature: 100°F

Amine sump (2 required)
Size: 8′ ID × 8′ high
Operating pressure: atmospheric
Operating temperautre: 160°F

Sand filters (4 required)
Size: 9′ ID × 15′ high
Operating pressure: 50 psig
Operating temperature: 185°F

Carbon filters (4 required)
Size: 9′ ID × 15′ high
Operating pressure: 50 psig
Operating temperature: 185°F

Lean amine cooler (2 required)
Heat load: 38.4 × 10^6 Btu/hr
U: 145
T_{LM}: 50°F
Area: 5,300 sq ft
Operating pressure: 270 psig

Semi-lean amine cooler (2 required)
Heat load: 156.9 × 10^6 Btu/hr
U: 145
T_{LM}: 50
Area: 21,640 sq ft
Operating pressure: 260 psig

Lean amine exchanger (2 required)
Heat load: 45.3 × 10^6 Btu/hr
U: 130
T_{LM}: 41°F
Area: 8,500 sq ft
Operating pressure: 50 psig

Semi-lean amine exchanger (2 required)
Heat load: 170.9 × 10^6 Btu/hr
U: 130
T_{LM}: 39°F
Area: 33,710 sq ft
Operating pressure: 60 psig

Regenerator condenser (2 required)
Heat load: 97.9 × 10^6 Btu/hr
U: 110
T_{LM}: 48°F
Area: 18,540 sq ft
Operating pressure: 50 psig

Regenerator reboiler (2 required)
Heat load: 303.5 × 10^6 Btu/hr
U: 120
T_{LM}: 50°F
Area: 50,580 sq ft
Operating pressure: 50 psig

Table 2.3 (*Continued*)

Amine reclaimer (2 required)
Heat load: 7.9×10^6 Btu/hr
U: 120
T_{LM}: 208
Area: 320 sq ft
Operating pressure: 50 psig

Lean amine pumps (3 required, including 1 spare)
Type: centrifugal
capacity: 1,475 gpm
Drive: motor
Hp: 325

Amine filter pump (2 required)
Type: centrifugal
Capacity: 620 gpm
Drive: motor
Hp: 25

Semi-lean amine pump (5 required, including 1 spare)
Type: centrifugal
Capacity: 2,640 gpm
Drive: motor
Hp: 900

Regenerator reflux pump (3 required, including 1 spare)
Type: centrifugal
Capacity: 265 gpm
Drive: motor
Hp: 15

Amine sump pump (3 required, including 1 spare)
Type: centrifugal
Capacity: 75 gpm
Drive: motor
Hp: 5

Amine makeup pump (2 required)
Type: centrifugal
Capacity: 65 gpm
Drive: motor
Hp: 3

Strong caustic pump (3 required, including 1 spare)
Type: centrifugal
Capacity: 80 gpm
Drive: motor
Hp: 5

Weak caustic pump (2 required)
Type: centrifugal
Capacity: 80 gpm
Drive: motor
Hp: 5

Caustic makeup pump (8 required, including 1 spare)
Type: centrifugal
Capacity: 40 gpm
Drive: motor
Hp: 1/4

Amine surge tank (2 required)
Size: 40′ ID × 20′ high
Operating pressure: atmospheric
Operating temperature: 210°F

Amine makeup tank (2 required)
Size: 15′ ID × 16′ high
Operating pressure: atmospheric
Operating temperature: 105°F

Caustic makeup tank (2 required)
Size: 10′ ID × 12′ high
Operating pressure: atmospheric
Operating temperature: 105°F

Gas filter separator (2 required)
Capacity: 226.1 MM scfd

Antifoam injector (2 required)
Type: piston-plunger
Capacity: 5 gpm
Drive: motor
Hp: 1/4

The flowsheet and the equipment list contain the basic background information required prior to establishing equipment costs. The flowsheet also serves as a means of determining the utility requirements such as steam, electricity, process and cooling water, and inert gas.

A factored study estimate also requires preliminary engineering (equipment sizing) and at least a cursory investigation of the utility and facility support systems. An overall understanding of the project may not be required to do capital estimating, but it certainly eliminates many potential pitfalls and problems and gives the estimator a better feel for assigning the factors to be used.

In a factored estimate it is of paramount importance that the bare equipment costs be determined as accurately as possible because the entire estimating system is based on the bare equipment costs. It is important to remember that in most chemical and petrochemical plants the bare equipment costs accounts for about 18–35% of the total plant capital requirement. Therefore, an error made in the bare equipment cost determination will be amplified by a factor of 3–5 to 1 in the overall estimate.

A piece of equipment is required to receive, hold, pump, compress, and release material. Some equipment can be identified as "off-the-shelf items." These are manufactured in large quantities and are readily available since the demand for such items is high. Included in this category are pumps, compressors, heat exchangers, and crushing and grinding equipment. Other items are especially designed specifically for a particular application, as in the case of a new or developing process, and thus must be manufactured or fabricated as needed.

The cost of equipment can be obtained from the following:

1. Firm bids and quotations
2. Previous project equipment costs
3. Published equipment cost data
4. Preliminary vendor quotations
5. Scale-up of data for similar equipment of other capacities

With basic design data on hand, the specifications for the equipment can be prepared. These specifications can be submitted to vendors for quotations. The quotation may be firm or preliminary and may also include a delivered price from the vendor's location to the site of the proposed project.

Previous project equipment costs may be available in the files of the firm interested in building the plant. Use of such data has the advantage of maintaining company confidentiality, which otherwise is lost when quotations are solicited. Care, however, must be taken in using such data to take into account the time that the data were originally obtained.

The installed cost of equipment includes the following:

1. Cost of purchasing or fabricating the equipment
2. Cost of transportation to the plant site
3. Cost of installing the equipment
4. Cost of providing foundations, piping, electrical connections, etc.

There are several approaches to determining the total installed costs. The most accurate is a detailed determination of every operation involved, including designs detailing requirements for foundations, piping, supports, electrical connections, painting, insulation, instrumentation, or in essence putting the equipment into operational condition.

However, "shortcut" factored methods are available, and in many cases are quite suitable for preliminary studies. It must be recognized that these shortcut methods are derived from prior collection of the detailed costs of purchasing, installing, and putting into operation similar equipment.

The factors used to "build up" an estimate have been the subject of many papers. The only way to develop such factors is from historical data and/or from experienced estimators and engineers. Many publications list ranges for such factors, but the ranges are so wide as to be of little practical use.

One technique that has been available for many years is the use of Lang factors.* These factors take into account the type of operation involved in a plant:

$3.10 \times$ delivered cost of equipment for solid processing plant

$3.63 \times$ delivered cost of equipment for solid-fluid plant

$4.74 \times$ delivered cost of equipment for fluid process plant

For example, in a coal preparation plant (a solid processing plant) where delivered equipment costs are estimated to amount to $10,000,000, the total cost would be about $31,000,000. For an oil-shale-to-crude-oil plant (a solid-

* H. J. Lang, "Simplified Approach to Preliminary Cost Estimates," *Chemical Engineering*, *55*, *112–113* (June 1948).

Table 2.4 Distributive Percentage Factors*

Type of system:		Solids		Solids-gas				Gas processes				Liquid and liquid-solid	
Temperature:		≤400°F	>400°F	≤400°F		>400°F		≤400°F		>400°F			
Pressure:				≤150 psig	>150 psig	≤150 psig	>150 psig	≤150 psig	>150 psig	≤150 psig	>150 psig	≤150 psig	>150 psig
Foundations	Material[a]	4	5	5	6	6	6	5	6	6	5	5	6
	Labor[b]	133	133	133	133	133	133	133	133	133	133	133	133
Structures	Material	4	2	4	4	5	6	5	5	5	6	4	5
	Labor	50	100	100	100	50	50	50	50	50	50	50	50
Buildings	Material	2	2	2	2	5	4	3	3	3	4	3	3
	Labor	100	100	100	50	50	100	100	100	100	100	100	100
Insulation	Material	–	1.5	1	1	2	2	1	1	2	3	1	3
	Labor	–	150	150	150	150	150	150	150	150	150	150	150

Instruments	Material	6	6	2	7	7	8	6	7	7	7	6	7
	Labor	40	40	40	40	40	75	40	40	75	40	40	40
Electrical	Material	9	9	6	8	7	8	8	9	6	9	8	9
	Labor	75	75	75	75	75	75	75	75	40	75	75	75
Piping	Material	5	5	35	40	40	40	45	40	40	40	30	35
	Labor	50	50	50	50	50	50	50	50	50	50	50	50
Painting	Material	0.5	0.5	0.5	0.5	0.5	0.5	0.5	0.5	0.5	0.5	0.5	0.5
	Labor	300	300	300	300	300	300	300	300	300	300	300	300
Miscell-aneous	Material	3	4	3.5	4	4	4.5	3	4	4	5	4	5
	Labor	80	80	80	80	80	80	80	80	80	80	80	80
	Material[a]	33.5	35.0	59.0	72.5	76.5	79.0	76.5	75.5	73.5	79.5	61.5	73.5
Totals:	Labor[a]	24.87	29.25	41.25	47.98	48.73	55.08	48.45	49.23	48.83	53.20	41.25	50.53

* These factors have been adopted by AACE International as a part of their Recommended Practice No. 16R-90, "Conducting Technical and Economic Evaluations in the Process and Utility Industry."

[a] Bare equipment cost × factor.

[b] Material cost × factor.

fluid plant) where equipment costs amount to $75,000,000, the total cost would be about $270,000,000. For an oil refinery (a fluid process plant) the total cost would be about $76,000,000 where the equipment costs are estimated to be $16,000,000.

Solids-handling systems generally are not enclosed and are operated at ambient conditions. Fluid systems are enclosed, often operate at elevated pressure and/or temperature, and present more complex control requirements, resulting in higher costs than those for solids-handling systems. Thus the Lang factor for solid-fluid systems is larger than for solids-only systems. Fluids-only systems similarly have a larger factor than that of solids-fluids systems.

A more complex factored estimating system is one in which all items involved in installing equipment and placing it into operable condition are factored. These factors are called "distributive percentage factors." Table 2.4 lists such factors for four different plant types: (1) solids, (2) solids-gas, (3) gas processes, (4) liquid and liquid-solids. Temperature and pressure are also taken into account. The break point for temperature is 400°F and for pressure it is 150 psig. All major items required for the complete installation are considered. The bare equipment cost is used as the base for the calculations involved. The percentage factor is applied to establish the installation material cost. Then the installation material cost is used as a base for determining the labor cost involved.

As an example, a gas-to-gas heat exchanger has a delivered price of $10,000 and is designed to operate at 800°F and at a pressure of 125 psi. Table 2.5 illustrates the use of these factors for putting in the heat exchanger in an operable mode.

To install any type of equipment, provision must be made for the items included in Table 2.4. However, the labor cost of physically installing the unit must also be determined. Table 2.6 lists the labor factors involved in installation of various types of equipment. These factors were developed from a series of detailed estimates and represent average values. For items not listed in Table 2.6, the following rules of thumb may be used in the absence of data to the contrary:

1. Equipment such as hoppers, chutes, etc., (no moving parts) requires a setting cost of about 10% of the bare equipment purchased cost.
2. Rotary equipment such as compressors, pumps, fans, etc., requires a setting cost of about 25% of the bare equipment purchased cost.
3. Simple machinery such as conveyors, feeders, etc., requires a setting cost of about 15% of bare equipment purchased cost.

Table 2.5 Typical Costs for Placing Heat Exchanger in Operable Modc (Bare Equipment Cost = $10,000)

	Material	Labor
Foundations	$ 600	$ 798
Structures	500	250
Building	300	300
Insulation	200	300
Instruments	700	525
Electrical	600	240
Piping	4000	2000
Painting	50	150
Miscellaneous	400	320
Total:	$7350	$4883

The total cost of installing a piece of equipment, using this estimating technique, then equals the bare equipment cost *plus* the installation labor cost *plus* the materials and labor costs as determined using the distributive percentage factors. Thus in the heat exchanger example (Table 2.5) the total installed cost equals $10,000 (bare cost) + $10,000 × 0.20 (installation labor cost) + $7,350 (materials cost) + $4,883 (labor cost), for a total of $24,233. Because the various factors are not precise, the final answer should be rounded to no more than three and preferably two significant figures to avoid implying a higher level of accuracy than actually exists. In this example, the estimated cost of the heat exchanger, therefore, should be reported as $24,000 rather than $24,233.

Table 2.4 is intended to set some guidelines for determination of both materials and field labor associated with bulk accounts. The material-labor split is important if any attempt is made to estimate field labor requirements. Using the first numerical column as an example, the "4" indicates 4% of the bare equipment cost as the factor for foundation material (concrete, rebar, etc.). The "133" indicates that 133% of the above 4% should be used as the labor cost to install the foundations or a total percentage of 9.32% of the bare equipment costs for foundations. As is indicated at the top of the column, this is for a solids handling system. This represents only one methodology

Table 2.6 Percentage Labor Factors for Bare Equipment Installation

Equipment type	Factor	Equipment type	Factor
Absorber	20	Hammermill	25
Ammonia still	20	Heater	20
		Heat exchanger	20
Bailey feeder	25	Hopper	15
Ball mill	30		
Bin	15	Knockout drum	15
Blower	35	Lime leg	15
Briquetting machine (with mixers)	25		
		Methanator (catalytic)	30
Centrifuge	20	Mixer	20
Clarifier	15		
Coke cutter	15	Precipitator	25
Coke drum	15	Pump	30
Compressor	35		
Condenser	20	Regenerator (packed)	20
Conditioner	20	Retort	30
Conveyor	25	Rotoclone	25
Cooler	20		
Crusher	30	Screen	20
Cyclone	20	Scrubber (water)	15
		Settler	15
Decanter	15	Shift converter (catalytic)	25
Distillation column	30		
		Splitter	15
Evaporator	20	Storage tank	20
		Stripper	20
Fan	25		
Feeder (vibratory)	25	Tank	20
Filter	15		
Fractionator	25	Vaporizer	20
Furnace	30	Vibratory feeder	25
Gasifier	30	Water scrubber	20

of estimating labor; for example, workhours per yard of concrete times an appropriate labor rate could well be used for the labor component. Sometimes the factors used include both the materials and labor; however, treating materials and labor separately allows the estimator to make an additional check on the reasonableness of the estimate. Also, an estimate of total labor workhours can be made from the plant summary (discussed later).

Other important cost considerations in a factored estimate are the workhours allowed for setting the "bare equipment," the field indirects, engineering, overhead and administration, a contingency, and a contractor's fee. Each of these will be discussed separately.

Workhours to set equipment are always derived from historical data and/or from the experience of the engineers and estimators on the project. Engineering and construction firms maintain workhour tables for setting different types of equipment according to weight, horsepower (rotating equipment), and so on. Percentage factors such as those given in Table 2.6 may also be used, but vary widely from 5% to 35% of the bare equipment cost depending on the difficulty of the work. Although in the overall estimate this allowance is rarely an overriding consideration, these costs should not be omitted.

Field indirects are often factored from the field direct labor costs and cover such items as temporary construction services including offices, warehouses, working areas, roads, walks, fences, parking areas, unloading facilities, power, light, telephone service, minor construction, general site drainage, sanitary facilities, and any other utility systems required. Field indirects also include maintenance of tools and equipment, materials handling, supplies, equipment rental and/or purchase, fuels, and lubricants.

Field office costs include nonmanual supervision, administration, warehousing and purchasing, first aid, safety and medical, nonmanual payroll adds and benefits, and field office overhead.

Home office costs included in field indirects are project management, technical services (project controls), construction management, procurement, and commercial services.

The factors for the above items will vary from 40 to 100% of the direct field labor and are somewhat dependent upon what is included or excluded from the above list. The most common factors are between 60% and 75% of the total field labor.

Engineering is somewhat self-explanatory and is the cost of engineering the plant. It ranges from 2 to 10% of the plant cost including field indirects. Obviously, the cost depends upon the amount of new engineering required compared to pre-engineered units and the complexity of the project.

Table 2.7 Capital Cost Summary

Item	Quantity	Cost (dollars)		Total cost (dollars)
		Material	Labor	
1.				
2. (Individual equipment items)				
3.				
4.				
5.				See labor factors for bare equipment Table 2.6
....				
....				
....				
....				
....				
....				
Subtotals		Line 1-A	Line 1-B	Line 1
Foundations				
Structures		Base on Line 1-A and Table 2.3		
Buildings				
Insulation				
Instrumental				
Electrical				
Piping				
Painting				
Miscellaneous				
Total direct		Line 2-A	Line 2-B	Line 2

Table 2.7 (*Continued*)

	Total cost (dollars)
Field indirect (% of total direct labor).	Line 3
Total construction (lines 2 & 3).	Line 4
Engineering (% of line 4)	Line 5
Overhead and administration (% of line 4)	Line 6
Line 7 (lines 4, 5, and 6)	Line 7
Contingency (% of line 7)	Line 8
Line 9 (lines 7 and 8)	Line 9
Fee (% of line 9)	Line 10
Total investment, line 11 (lines 9 and 10).	Line 11

Table 2.8 Typical Capital Cost Summary

Date: 08/06/xx	Equipment and Installation
By: P. Wellman	Title: ABC Alcohol Company
Report: Ethanol	Unit: Fermentation

Item	Quantity	Cost (dollars)		Total cost (dollars)
		Material	Labor	
Fermenter	8	904,800	90,500	
Fermenter agitator	8	112,000	11,200	
Fermenter cooler	4	519,200	51,900	
Fermenter circ. pump	8	170,400	25,600	
Fermenter cleaner	8	16,000	1,600	
Beer well	1	195,800	19,600	
Beer well agitator	2	28,000	2,800	
Beer well cleaner	1	3,000	300	
Sterilant scale	1	1,400	100	
Sterilant pump	1	1,100	200	
Sterilant tank	1	14,500	1,500	
Sterilization pump	1	18,500	2,800	
Sterilant tank agitator	1	1,300	100	
Distillation feed pump	2	44,000	6,700	
CO_2 Offgas scrubber	1	55,400	5,500	
Scrubber pump	1	3,200	500	
Scrubber blower	1	30,000	4,500	
Scrubber chiller	1	30,000	3,000	
		2,148,600	228,400	2,377,000
Foundations		107,400	142,800	
Structures		85,900	43,000	
Buildings		64,500	64,500	
Insulation		21,500	32,300	

Table 2.8 (*Continued*)

Item	Quantity	Cost (dollars) Material	Cost (dollars) Labor	Total cost (dollars)
Instrumentation		128,900	51,600	
Electrical		171,900	128,900	
Piping		644,600	322,300	
Painting		1,100	3,300	
Miscellaneous		85,900	68,700	
		1,311,700	857,400	2,169,100
Total direct		3,460,300	1,085,800	4,546,100
Field indirect @ 50% of labor				549,900
Total construction				5,096,000
Engineering @ 5%				254,800
Overhead and administration @ 5%				254,800
				5,605,600
Contingency[a]				
				5,605,600
Fee @ 5%				280,300
Total capital cost				5,886,900
				≈ $5,900,000

[a]Contingency taken on total plant.

Overhead and administration, as used here, is the cost to the owner for administering and managing the contract. The costs vary widely depending on company or owner practice. The costs range from as little as 1 to as much as 10% of the plant costs and include frontline supervisor and operator training. Furthermore, it appears that the practice of more complete management and cost control by owners has become more prevalent. Major construction projects are no longer let on a cost-plus basis without owner management and control as was the normal practice for many companies in the past.

Contingency is a much maligned item in a capital estimate. The allowance is *not* intended to cover scope changes or to hide poor estimating but is intended to reflect the accuracy of the estimate considering the stage of development of the project. The contingency is intended to cover omissions and unforeseen expenses caused by the lack of complete engineering and, therefore, an inadequate scope of work. The estimate including contingency should give a 50:50 probability of overrun. The contingency for a factored estimate should be in the 20–30% range. The actual determination of contingency will be discussed later.

The fee is simply the contractor's pay for the work.

Table 2.7 summarizes the method described above and Table 2.8 gives an example of its application for the fermentation section of an ethanol plant.

The distributive percentage factor approach has considerable flexibility compared to a purely arithmetic approach. It allows comparisons of various factors with historical data or other sources of information and an estimate of total construction labor workhours may be made if desired.

The preceding procedure is used to estimate the cost of the battery limits of each of the processing units. Therefore, only part of the capital requirement has been determined. Items not covered by the above include "off-sites" and "support systems" such as the steam plant, power plant (if any), process water supply, potable water, sewers and drains, product storage and loadout facilities, power distribution, interconnecting piping, initial chemicals, fire water loop, communication system, plant and instrument air, wastewater treatment, administrative buildings, parking lots, roads, and fences. This is not intended to be an exhaustive list but does indicate the magnitude of the costs.

Although all of the items above may be costed using factors, factors are rarely used as the only basis for the numbers.

A preliminary utility balance is normally done for a serious attempt to determine capital investment. Table 2.9 is an example of a form that may be used to facilitate such a balance and also to summarize the equipment costs. When properly used, a form of the type illustrated in Table 2.9 affords the estimator a simple straightforward way to generate the information necessary to complete a preliminary utilities balance. Also, the form provides for the buildup of catalyst and chemical requirements, both initial and annual. The steam requirement and/or production (heat recovery) can be recorded in pounds per hour as well as the quality of the steam (pressure and temperature). The "remarks" area provides for recording of back up calculations such as steam requirements for a turbine, etc. Thus, Table 2.9 gives an overall view of one processing entity, that is, the *battery limits* capital requirement and the information required to proceed with the estimate.

The method described above is used to determine the capital investment for all units directly involved in the production of a product or products. Such estimates are known as *battery limits* estimates. They are estimates of required investment for a specific job without regard to required supporting facilities which are assumed to exist already as would be true in the case of an addition to an existing plant.

Many estimates, however, are for totally new facilities on an undeveloped site. In such cases complete site development must be included in the estimate along with required off-site supporting facilities which are needed to make the plant self-sufficient. A complete estimate of this nature is known as a *grass roots* or *green field* estimate, and may include units to produce steam, electricity, treated water, packaging and shipping facilities, laboratory and office buildings, cafeterias and medical facilities, railroad spurs and roads, firefighting and safety equipment, garage and maintenance shops, etc.

If the estimate is a complete grass roots estimate, the offsites, utilities, and support facilities can be and sometimes are factored. The information derived from preliminary utility balances gives a substantial basis for quantifying such factors. There are also tables of factors in the literature for the utilities, facilities, and offsites to support a processing plant. However, the major utilities, facilities, and unusual items necessary to support the processing plant need at least cursory investigation and/or a separate estimate to maintain desired accuracy. Some of the items that must be scrutinized are the electric distribution system, interconnecting piping, the steam plant, power generation (if generated on site), wastewater treatment, and any unusual support facility. Many of these items are discussed further in Chapter 4.

Land costs and interest during construction may also be included in the total capital investment determination. However, land is not depreciable since the land may be resold and reused for other purposes after the plant is no longer in operation. Conversely, interest during construction is generally capitalized and depreciated.

Table 2.9 Equipment and Utility Cost Summary

DATE: 08/06/xx
BY: P. Wellman
REPORT: ABD Alcohol

Equipment Cost Summary
TITLE:
UNIT: Feed Storage and Fermentation

ITEM	Quantity	Material, $	Labor, $	Total cost, $	Catalyst and Chemicals		Cooling water gpm	Hp required		KW required Total	Steam required		Steam produced	
					Initial	Annual		Unit	Total unit		lb/hr	Level	lb/hr	Level
Yeast mix tank and mixer	1							5	5					
Fermentation tank mixer	8							25	200					
Beer well tank and mixer	1							10	10					
Sterilant tank and mixer	1							2	2					
Fermentation circ. pump	8							100	800					
Sterilant weigh scale	1													
Fermentation cooker	8						6000							
Centrifuge	1							250	250					
Recycle filtrate pump	1							75	75					
Yeast slurry pump	1							1	1					
Sterilization pump	1							1	1					

Distillation feed pump	2							50	100					
CO_2 offgas scrubber	1													
CO_2 blower	1							25	25					
CO_2 scrubber chiller	1						500	50	50					
Totals							6500		1519	1200				

Foundations Structures Buildings Insulation Instrumentation Electrical Piping Painting Miscellaneous			
Totals			
Total direct			
Field indirect			
Total construction			
Engineering Overhead and administration			
Subtotal			
Contingency			
Subtotal			
Fee			
Total capital cost			

Remarks

Power factor considered; factor = .746/95 = .785

Initial catalyst and chemicals may also be included. The initial fill in vessels required for full operation becomes a capital investment requirement. Subsequent refills or replacement costs are included in operating costs.

Working capital is another significant investment requirement. It is needed to meet the daily or weekly costs of labor; maintenance; purchase, storage, and inventory of field materials; and storage and sales of product(s). Stated another way, working capital includes the following:

1. Inventories
2. Accounts receivable
3. Cash for wages and materials
4. Other accounts payable

Normally, working capital accounts for one month's allowance for the four items, but in some cases, it may range as high as three months' allowance. The working capital allowances will depend on many factors. Several of these factors which must be considered are the following:

1. Seasonal demand for product—for example, fertilizers
2. Availability of certain fuels such as coal which are produced on the average of 230 days/yr and may be required for a 330-day operation
3. Delays in shipping due to strikes or weather conditions and resultant delays in invoicing

Another cost item that should be considered in determining total capital investment is start-up cost. A major fault that can occur during the initial operation of the plant is the possibility that the product(s) may not meet specifications required by the market. As a result equipment changes—modifications or additions—may be necessary or additional equipment may be required to meet full design goals. Start-up can add 10–20% to capital investment. The low value applies to a plant of a type that has been built and operated previously. The 20% figure would apply to new or novel plant installations. Start-up costs are discussed in more detail in Chapter 11.

Table 2.10 shows a capital summary for a factored type estimate (excluding land and start-up costs). From the table the deficiencies of a factored estimate are obvious; that is, any individual entry could be off by as much as 50% or more. It is presumed all too often in factoring that some costs are high and some are low tending to cancel each other out. Historically, how-

Table 2.10 Total Estimated Capital Rquirements and Operating Personnel

	ABC Gas Company, Synthesis Gas		
Unit	Cost Millions of dollars	Percent	Estimated number of operators per shift
Coal receiving & storage	4.034	2.29	2
Slurry preparation	5.541	3.15	2
Gasification	22.990	13.07	4
Particulate removal and waste heat recovery	3.157	1.79	1
Acid gas scrubbing	37.467	21.29	3
Water gas shift	13.346	7.58	2
Sulfur recovery	8.126	4.62	2
Steam plant	6.225	3.54	2
Plant facilities	7.566	4.30	2
Plant utilities	10.845	6.17	2
Total construction	119.297	67.80	22
Initial catalyst requirements	3.012	1.71	
Total plant cost (insurance and tax bases)	122.309	69.51	
Interest during construction	11.007	6.26	
Subtotal for depreciation	133.316	75.77	
Contingency	33.329	18.94	
Working capital	9.314	5.29	
Total investment	175.959	100.00	

ever, factored estimates are low more often than high with an expected accuracy of +30 to −15% overall.

PRELIMINARY (BUDGET AUTHORIZATION) ESTIMATES

The preliminary or budget authorization estimate requires all of the items specified for a factored estimate plus more detailed engineering work. The estimate should be based on complete or nearly complete process engineering. The plant engineering should be 2–5% complete and the project must be site specific with preliminary soils work completed. A plot plan including pipe-rack routes, storage areas, warehouse, shops, etc., should have been completed.

In addition to flowsheets and materials and energy balances, at least preliminary piping and instrument diagrams (P&ID's) should be completed. The P&IDs should give line sizes and identify control loops, valves, pressure relief, etc., which allows a rough piping takeoff or a much improved basis for estimating both the piping and the instrumentation factors. Also, a preliminary single-line electrical drawing showing the substation, transformers, circuit breakers, distribution lines, motor control centers, and so forth, must be completed.

The major difference between a preliminary (budget authorization estimate) and the factored estimate is not necessarily procedure but rather the information available upon which to base the factors or to make a crude take-off of bulk accounts. Process equipment is more defined, with at least vendor telephone quotes on all major equipment and significantly more information concerning the site, foundations, drainage, and other considerations affected by site selection.

The basis for developing offsites, utilities, and facilities requirements should be improved considering the additional engineering referred to above. Again, the estimating may be done in-house or in cooperation with vendors; however, using either method the accuracy of the estimates should be much improved over a purely factored estimate.

Forms similar to those used to develop the factored estimate may be used; however, each owner and/or engineering/construction contractor has developed in-house forms and procedures which will be discussed later.

It is reasonable to expect accuracy of −10 to +20% for a preliminary estimate which has been well executed. As stated previously, a careful examination of the elements which are used to build the estimate is necessary to make an intelligent assessment of accuracy.

DEFINITIVE (PROJECT CONTROL) ESTIMATES

A definitive estimate is made when plant engineering is 40–45% complete and/or quotations have been received on 90–95% of the major equipment. The engineering includes, in addition to items listed earlier, complete plot plans and elevations, final P&IDs, equipment data sheets, sketches of major foundations, complete specifications, building sketches, structural drawings, offsite definition and design, and the final utility balance.

The definitive estimate requires the development of a complete code of accounts (see Chapters 3 and 4) and construction schedule, including estimates of labor loading and indicating the construction milestones and their expected dates of completion, the last being mechanical completion.

The bulk accounts are based on takeoffs from the drawings, and the field indirects are based on actual projected field needs for equipment rental or purchase, labor, temporary facilities, utilities, expendable tools, and supplies.

The necessary home office staff and support groups are specified and their costs estimated based on actual salaries, anticipated travel, and technical support.

The engineering workhours have been estimated based on the number of drawings required, salaries, and overheads. The fee should have been determined by the contractual arrangements between the owner and the contractor.

The accuracy of the definitive estimate should be between –5 and +15%.

DETAILED (FINAL) ESTIMATES

The detailed estimate is initiated at the completion of engineering and, thus, all engineering drawings including isometrics are done. The detailed estimate is never factored but based on actual equipment cost, bulk accounts, labor projections, and usually a scaled model of the plant. Any inaccuracies are normally caused by estimating errors and/or labor productivity problems.

The accuracy of such an estimate should be ±5%.

During the normal evolution of a project, all of the foregoing types of estimates could be completed, but this is rarely the case. The estimate is a dynamic tool which should be continually updated and corrected as the project progresses. Thus, the estimate moves from a screening ratio estimate to a detailed estimate for a new installation without a finite determination of type at any particular time. In fact, most of the time the estimate is a mixture of all or part of the generalized types listed. This fact should be considered when the contingency determination and accuracy assignments are made.

PROBLEMS AND QUESTIONS

2.1 The cost of a 200 sq. ft., shell-and-tube, floating head, carbon steel heat exchanger installed in 1990 was $105,000. What would the cost of a heat exchanger twice as large have been in 1994? Hint: The *Chemical Engineering* index had a value of 357.6 in 1990 and 362 when the larger heat exchanger was purchased. The *Marshall & Swift Index* had a 1990 value of 915.1 and 970 in 1994.

2.2 What data should be shown on a factored estimate flowsheet?

2.3 a) The bare equipment cost for a coal preparation plant is $4,800,000. What is the approximate cost of the total installation? b) What would the total cost be if the equipment in (a) was for an addition to an oil refinery?

2.4 Given the following list of bare equipment costs for a factored estimate:

	Cost/M dollars	
	Material	Installation labor
Fermenter	800.1	80.0
Fermenter agitator	120.2	12.0
Fermenter circulating pump	191.0	28.7
Beer well	201.2	20.1
Distillation feed pump	51.6	7.7
CO_2 offgas scrubber	59.4	5.9
Scrubber blower	32.0	6.7
Scrubber chiller	32.0	3.0

What would the total capital cost be for a battery limits installation? The plant operates at low pressure.

2.5 a) What are the five classifications of estimates discussed in Chapter 2? b) What are the ranges of anticipated accuracy for each classification?

2.6 a) What is the purpose of a contingency factor in a capital estimate? b) List items not covered by the contingency.

2.7 a) Define *working capital.* b) List the four elements included in working capital.

2.8 Why is a utility balance important in preliminary estimates?

2.9 What capital cost items are excluded from a battery limits estimate?

2.10 What should the probability of overrun be on a factored capital estimate including the contingency?

2.11 Why is an allowance made for startup costs?

2.12 a) How much of the plant and process engineering should have been completed to do a preliminary (budget) capital estimate? b) Why are P&IDs so important for a preliminary estimate?

2.13 What data are required to do a final detailed estimate?

2.14 a) What are cost indexes used for? b) What does the Marshall and Swift (M&S) cost index cover? c) What do the two *Engineering News-Record* (ENR) cost indexes cover?

2.15 In scaling up equipment costs, what formula is used? Is there a suggested limit of scale-up for use of this formula (2:1, 5:1, 10:1, or no limit)? Discuss.

2.16 What formula is used for time-dating cost data, i.e., for adjusting old data for inflation? Is there a suggested time limit for use of this formula (2 years, 5 years, 10 years, or no limit)? Discuss.

2.17 Equipment for a new plant is estimated to cost $6,000,000 delivered. Installation labor cost is estimated at $2,000,000. The plant will produce petrochemicals from petroleum feedstocks at high pressure (approximately 300 psig). Estimate the total cost of constructing the plant, assuming:

Field indirects at 50% of direct labor,
Engineering at 12.5% of base construction costs,
Overhead and administrative cost at 7.5% of base construction cost,

Assume further that a 10% contingency is adequate and that the contractors fee is 7.5%.

2.18 You estimate that the total cost of building a new plant will be $10,000,000 including land. Annual operating costs will be $8,000,000. Your company can raise $11,500,000 in investment capital without overextending itself financially. The plant is expected to be very profitable. Should it be built? Why or why not?

2.19 Estimate the total installed cost of a 10,000 cu. ft/minute cyclone dust collector. In 1990 the purchase price of such a unit was $12,000. The

Marshall & Swift Equipment Cost Index had a value of 915 in 1990 versus about 970 now.

2.20 You wish to estimate the cost of constructing a structural steel silo with a capacity of 12,000 cubic feet. Past project records indicate:

a. a 6,000 cubic-foot silo of a similar type cost $500,000 installed in 1990 (M&S Index = 915)

b. a 9,000 cubic-foot silo of a similar type cost $825,000 installed in 1992 (M&S Index - 945)

c. you estimate that the M&S Index will have a value of 980 when the new silo is built.

Assuming that design conditions in all three cases resulted in the silos having the same height, what is the estimated cost of the new silo?

3

Organizing the Capital Estimate

A conceptualization of the product (the estimate) and a methodical procedure for attaining the desired results are as necessary in capital estimating as in other endeavors.

A word of caution: *Do not be overwhelmed by the complexity or misled by the apparent simplicity of the work* as either approach can be catastrophic to achieving the desired result. Concerning complexity, the work must be broken down into manageable divisions and subdivisions. The statement on simplicity is intended to indicate a huge trap of oversimplification and thus inexcusable omissions from the capital cost estimate.

The two requirements for reasonable estimating are process flowsheets and a "statement of work" (scope). The preparation of process flowsheets requires materials and energy balances. It forces one to establish operating temperatures and pressures and to become familiar with the processing scheme and/or schemes. Processing almost always requires consideration of more than one process alternative and therefore requires tradeoffs to arrive at the final plant configuration. After the development of the flowsheet and/or flowsheets, a statement-of-work, sometimes called scope-of-work, should be written. The written scope often reveals omissions and/or illogical development of the flowsheets. The scope statement also serves as a constant reminder of what is to be accomplished and how it is to be done.

For an order-of-magnitude (ratio) estimate, flowsheets and scope-of-work statements are not prepared, but all other estimate classifications require the preparation of both.

As stated above, some procedure must be used to organize the estimate into manageable divisions of work, this may be accomplished by an approach such as considering unit operations for a small plant to using a full-blown code of accounts for a large complex installation. In any situation, regardless of complexity, the estimator does one thing at a time; thus the estimate is built as a brick wall, one piece at a time. Obviously, more than one person can work on the estimate but overall coordination of effort is required.

Most generally the flowsheets and offsite facilities are divided into sections by some sort of numbering and/or identification system. For example one might use a system such as is illustrated in the following table:

Plant	Name	Equipment
1	Raw materials handling	All equipment numbers start with 1
2	Plant feed system	All equipment numbers start with 2
3	Process plant 1	All equipment numbers start with 3
4	Process plant 2	All equipment numbers start with 4

Then an alphabetical character could be used to identify various types of equipment, such as V for vessel. It follows then that 1V-101 would be a vessel located in Plant 1 and 2V-101 would be a vessel located in Plant 2. It is of paramount importance to use only one numbering code for a project. The owner, contractor, and subcontractors must use the same code. Normally this is the system used by the prime contractor. More than one code of accounts on any one project is an absolute disaster because of the inevitable confusion and errors which occur when different coding systems are used. The above system can be used to tabulate the bare equipment cost for each section of the project. The sections should be carefully selected, grouping together similar types of equipment. The careful grouping of equipment facilitates the determination of installation factors in a factored estimate and should follow a logical processing sequence.

For general and heavy construction, most design professionals now utilize the MasterFormat coding system established by the Construction Specifications Institute. MasterFormat uses 16 divisions to classify codes of accounts. They are:

Division 1—General Requirements
Division 2—Sitework
Division 3—Concrete
Division 4—Masonry
Division 5—Metals
Division 6—Wood and Plastics
Division 7—Thermal and Moisture Protection
Division 8—Doors and Windows
Division 9—Finishes
Division 10—Specialties
Division 11—Equipment
Division 12—Furnishings
Division 13—Special Construction
Division 14—Conveying Systems
Division 15—Mechanical
Division 16—Electrical

As is obvious from this list, if MasterFormat is used in the project cost coding system, it can encompass the entire "grassroots" installation, including roads, buildings, etc. Process facilities and equipment, as discussed to this point, would generally be coded under Division 11. Chapter 4 discusses MasterFormat in more detail and provides numerous rules of thumb for estimating various items in the 16 divisions.

Determination of the bare equipment cost is the easy part of a cost estimate. Estimating the bulk accounts, the labor, and the management costs is much more subjective and illusive than equipment estimating. However, the other work should be completed in parallel with the equipment cost estimate. Since some design is required for estimating, an equipment list should be made and a buildup of utility requirements completed at the time the estimate is being done. This approach avoids duplication of work.

A buildup of utility requirements as a result of the design work even on a preliminary basis is relatively easy. But if the utility balance is not done in conjunction with the estimate, then it is necessary to go systematically through the calculations to determine the balance (a duplication of work and a difficult task). It is also easier to accumulate the costs of catalysts and chemicals as the estimate is being done. This should be done both for the initial requirement and for the annual cost. A form designed to aid in the buildup of both of the above was given in the previous chapter (see Table 2.9).

Codes are also required for the bulk accounts (concrete, structural steel, piping, instrumentation, insulation, process buildings [not administration buildings, shops, etc.], painting, electrical and any other special accounts); these codes help organize the estimate and are often designated by alphanumeric combinations. For example, if C was used to designate the concrete account, then 1C-101 would indicate a foundation in plant 1. This allows not only a buildup of costs for plant 1 but also tracking of the concrete account, if desired.

There are as many different codes of accounts as there are significant engineering and construction contractors, but all are systematic approaches aimed at assuring the inclusion of *all* costs. A code of accounts is absolutely necessary for cost control on any project.

The following is a list of items that should be included in a code of accounts:

Land:
- a. Survey
- b. Fees
- c. Property cost

Site development:
- a. Site clearing
- b. Grading
- c. Roads, access and onsite
- d. Walkways
- e. Railroads (if applicable)
- f. Excavation and fill
- g. Fences
- h. Parking lot
- i. Piers (if applicable)

Process buildings:

Auxiliary buildings:
- a. Administration
- b. Cafeteria
- c. Garage
- d. Warehouse
- e. Shops
- f. Guard house
- g. Change house

- h. Shipping office
- i. Laboratories

Process equipment:
- a. Listed individually and checked against process flow diagrams (PFDs) and piping and instrument diagrams (P&IDs)

Nonprocess equipment:
- a. Office furnishings
- b. Cafeteria equipment
- c. Safety and medical supplies and equipment
- d. Shop equipment
- e. Yard material handling equipment
- f. Laboratory equipment
- g. Garage equipment

Process appurtenances:
- a. Piping
- b. Pipe fittings
- c. Valves
- d. Pipe hangers
- e. Insulation
- f. Instruments
- g. Instrument panels
- h. Electrical

Civil:
- a. Foundations
- b. Structural steel
- c. Pipe racks
- d. Site preparation

Utilities:
- a. Boilers
- b. Incinerators or flare
- c. Solid waste disposal
- d. BFW treatment
- e. Electrical substation
- f. Plant air system
- g. Communication system
- h. Water intake
- i. Water treatment
- j. Cooling towers
- k. Water storage
- l. Sewers and drains

m. Impoundment basins

Yard distributables:

a. Interplant piping
b. Product shipping
c. Weigh scales, truck and/or rail

Miscellaneous:

a. Initial catalyst and chemicals
b. Spare parts and uninstalled spares
c. Equipment rental
d. Premium time (overtime and holidays)
e. Freight
f. Taxes, insurance, duties

Engineering cost:

a. Administrative
b. Process, project, and general engineering
c. Drafting
d. Procurement, expediting, and inspection (quality control)
e. Travel
f. Reproduction
g. Scale model
h. Outside engineering

Construction expenses:

a. Temporary facilities
b. Temporary utilities
c. Expendable tools and supplies
d. Accounting and timekeeping
e. Safety and medical
f. Security
g. Home office expenses

As one can easily discern from the foregoing list, an exhaustive code of accounts needs to be developed and used in organizing an estimate of a construction project.

Also, it is apparent that at the inception of a project, some of the information needed is unavailable. Thus, the classification of estimates discussed in the previous chapter really depends upon how far the project has evolved (i.e., how much engineering, planning, etc., has been done).

A complete code of accounts is not presented here, but several such systems have been developed by various contractors and owners. The important consideration in evaluating a code is that such a system provides for the charging of all costs evolving from a construction project.

The organization of the estimate also depends on the status of the project. But in any presentation of capital costs, a summary is the first tabulation presented. In the normal evolution of making an estimate, it is the last table completed. The summary may take several forms. For example, for a factored estimate the summary may look like the following:

Capital summary

	Investment
Plant 1	$
Plant 2	$
Plant 3	$
Offsites	$
Facilities	$
Utilities	$
Total direct costs	$
Initial catalyst and chemicals	$
Total plant cost (insurance and tax base)	$
Interest during construction	$
Subtotal	$
Contingency	$
Working capital	$
Total investment	$

In the above format, most of the costs are factored, not detailed; that is, field indirects are factored and charged to each unit; engineering is charged in the same manner; both the facilities and utilities are factored, although the factors may have some basis such as a preliminary utility balance; also the contractor's fee is charged to each unit. Therefore, the format is determined by the methods used to build the estimate. Generally no attempt is made in a factored estimate to separate equipment, materials, labor, and/or subcontracts. Neither the field indirects nor engineering is detailed sufficiently to allow breakdown of the costs into the various categories, but rather a percentage of capital is used (i.e., lump sum numbers).

	Equipment	Materials	Labor	Subcontracts	Total
Plant 1	$	$	$	$	$
Plant 2	$	$	$	$	$
Plant 3	$	$	$	$	$
Offsites by individual cost center	$	$	$	$	$
Utilities by individual item	$	$	$	$	$
Buildings	$	$	$	$	$
Etc.	$	$	$	$	$
Direct plant Cost	$	$	$	$	$

Initial Catalyst and Chemicals	$
Field Indirects	$
Sales Tax	$
Home Office and Engineering Costs	$
Spare Parts Inventory	$
Subtotal	$
Contingency	$
Subtotal	$
Contractor's Fee	$
Total Plant Cost	$
Working Capital	$
Total Investment	$

Figure 3.1 Capital summary, preliminary estimates.

Budget and/or detailed estimates are usually presented in a different manner because more engineering has been completed and thus, more details are available. The general format shown in Figure 3.1 might be used. This format gives considerably more detail than the previous one and reflects the advances made in engineering, planning, and other areas of the project. The breakdown of costs shows that equipment, materials, labor, and subcontracts associated with each cost center or plant have been estimated separately. The equipment costs are normally improved as the project evolves and are self-explanatory. The materials column includes all of the bulk accounts as follows:

Concrete and rebar
Electrical
Instrumentation
Piping and related items
Valves
Painting
Piling (if needed)
Process buildings

This format includes only the costs to the battery limits for each cost center or plant. The interconnecting pipe, electrical distribution, pipe rack, fire water loop, cooling water distribution, cooling towers, waste treatment, water treatment, steam distribution, communication system, plant air system, etc., are treated as cost centers and thus costed separately.

For a preliminary (budget authorization) estimate many of the items listed in the preceding paragraph are or probably can be factored. The factors used, however, should be improved because of the better definition of the work from the advanced engineering and planning. As the project evolves, the basis for estimating factors improves; thus the estimate accuracy is improved. When the engineering is about 40–45% complete, the preliminary (budget authorization) estimate is factored less and less until the factors are replaced with actual takeoffs, etc., and the budget estimate becomes the definitive estimate.

The labor column on the form includes the labor to set equipment and the labor required for the bulk accounts. The labor rates must be established as well as the estimated hours to do the work. Also, an estimate of the crafts needed, when they are needed, and for how long they are required, must be made to allow planning to proceed and to establish craft requirements for the project.

The subcontracts column on the form is the cost of the work subcontracted, generally by the prime contractor. A prime contractor is almost a necessity on a construction site for logistics and overall control. Some of the items generally subcontracted are painting, some or all of the electrical work, site preparation, specialty processes and plants, piling (if needed), and civil work. It is possible that any or all of the work for a plant or cost center could be subcontracted. The subcontractor(s) labor requirement should be considered in the projection of total labor for the project and thus included when investigating craft availability, site labor density, etc.

By adding the equipment, materials, labor, and subcontracts costs, the total cost associated with each plant and/or cost center is determined. The total cost of all the plants and cost centers results in the direct plant cost shown on Figure 3.1.

The costs listed on Figure 3.1 below the direct plant costs on through home office and engineering can be and sometimes are charged to each plant and/or cost center. However, the foregoing format is probably more commonly used. Again, as the estimate is developed, catalyst and chemicals should be calculated.

Field indirects evolve from factors to actual cost as the engineering and planning progresses until the number of people, rental equipment required, etc., is estimated.

Local and/or state sales taxes are site specific and must be determined for each project and require investigation of the local tax laws. Tables are available which list the tax rates in the states and give a national average of such taxes. These rates are often used for generic and preliminary estimates. As the project progresses to being site specific, then the actual tax rate is used.

Home office and engineering may be lumped together as shown on Figure 3.1; however, a separate determination of these costs is generally made. During the normal progression through the various stages of estimating on a project, these costs go from generic factors to improved factors and finally are estimated based upon actual staffing, workhours, support required, number of drawings, etc.

The spare parts inventory should be built up during the engineering design culminating in a spare parts list which is then costed out. Early in the project factors are normally used.

The contingency determination will be discussed later; however, the contingency allowance should be sufficient to give a 50:50 probability of cost overrun on the project.

The contractor's fee is negotiated as a part of the construction contract, and in the early estimates a factor is normally used ranging from 1–5% of the contract value. Larger percentages are generally paid on smaller contracts. There are many ways to negotiate the fee, and sometimes bonuses and penalties are incorporated to award or to penalize the contractor's performance. For example, if the contractor brings the project in under budget and on time, the contractor may share in the savings. But if the contractor overruns the budget and prolongs the schedule, penalties may apply. Negotiation of a fair and equitable contract fee arrangement is important for all concerned with a project since having a contractor fail to make a reasonable fee is detrimental to contract execution.

The working capital is the out-of-pocket expense required to operate the plant and is a nondepreciable capital requirement. It is normally anticipated to be recovered at the end of the plant operations without loss. Working capital includes inventory (product), raw materials, goods in process, accounts receivable, and cash for operational expenses. The working capital is factored early in a project, but it is detailed in a definitive estimate.

The foregoing table is backed up by the determination of the capital required for each cost center and/or plant. A plant is defined as a processing entity, and a cost center is defined as a support system. Examples are distillation (a process entity) and a plant air system (a support system). Exactly how the estimate is set up or defined is immaterial if indeed a system is set up to account for *all* the costs.

The offsites and support systems are treated the same as processing plants and sequentially numbered in a code of accounts; for example, plant 40 could be the plant air system and plant 35 the steam plant. Then referring in the previous numbering system, 40V-101 would be a vessel in plant 40, and 35V-101 would be a vessel located in plant 35, etc.

The backup calculations and tables which support the capital cost estimate for each plant and/or support system are the nitty gritty of estimating which many people try to eliminate, simplify, and/or ignore. These details however remain the basic building blocks for estimating. With the exception of ratio-type estimates, organized backup calculations are required and the methods used should reflect the status of the project.

For example, the following methods are used to cost out a pressure vessel such as a knockout drum depending upon the status of the project:

Order-of-magnitude estimate (rarely used for single items of equipment):

$$C_x = C_k \left(\frac{E_x}{E_k} \right)^n$$

where

C_x = Cost of drum

C_k = Cost of drum of size E_k

E_x = Size of new drum (E could be expressed in lbs.)

E_k = Size of known drum

n = Exponent, normally fractional

Study (factored) estimate: Vessel size determined using one of several methods such as gas velocity (at conditions), liquid residence time, superficial gas velocity, or any other suitable method

where

H = vessel height

D = vessel diameter

L = vessel length

For vertical drum $H = 2$ to $2.5 \times D$

For horizontal drum $L = 2.5$ to $3 \times D$

Vessel thickness can be calculated using the well-known formula

$$t = \frac{(P)\,(R)}{(S)\,(E) - 0.6P}$$

where

t = Shell thickness, in.

P = Design pressure, psi

S = Design allowable stress, psi

E = Efficiency of shell closure

R = Shell radius, in.

Calculate approximate weight of vessel
Cost out in $ per pound (cost curves)
Check against historical data if available

Preliminary (budget authorization) estimate:
- Vessel size has been determined
- Weight calculated
- Sometimes a preliminary data sheet, openings, etc.
- Metallurgy defined
- Costed in same manner
- Check against historical data
- Vendor estimate by telephone

Definitive (project control) estimate:
- Specifications written
- Data sheet available, metallurgy, etc.
- Inlet and outlet lines sizes known
- Weight of vessel known
- Vendor quotation
- Check against any other cost information available

Detailed (final) estimate:
- All items listed under definitive estimate
- Isometrics completed
- Plant model available
- Vendor quotation
- Always check against any available information

From the above, the first estimate could be exactly right but is certainly less likely to be right than the final estimate. During the development of the project the equipment cost estimates become more and more definitive, and therefore more accurate.

The accuracy also improves for the bulk accounts much in the same manner as for the equipment. That is, the definition improves, the drawings are completed, piping and instrument drawings completed, electrical diagrams completed, foundations designed, structural support identified, workhours to set equipment estimated, and workhours for the bulk accounts estimated based on the actual takeoffs. Probably the most difficult cost to evaluate is labor, which depends on productivity and is beyond the control of the estimator. Labor costs are often inflated by poor planning, bad supervision, restrictive union work rules, absenteeism, etc. These difficult-to-evaluate semi-social influences on labor costs are the cause of most of the inaccuracies in a final estimate.

A backup summation for a plant or cost center would look somewhat the same as the one given in the previous chapter but would be supported by much more detail in later estimates. That is, the workhours to set equipment would be the basis for the dollars charged, not a factor. The bulk accounts materials would be estimated from takeoffs, and workhours estimated for each account, the subcontracts would be let and listed separately, etc. It is most important to retain an organized record of the derivation of the estimate which may be ten times the amount of paper of the formal report. The organized backup and final cost record adds to the historical cost information and supplies the estimator with all information necessary to defend the estimate and/or estimating procedures. The backup material may also be used to determine why an estimate went awry, if it did. This can be used to correct procedures to avoid future errors.

How one organizes the backup calculations is a matter of personal preference. However, the faintest written record is better than the greatest memory and is essential to the development and improvement of a capital cost estimating system.

QUESTIONS

3.1 What is the word of caution given at the beginning of the chapter?

3.2 What are the two requirements for reasonable estimating?

3.3 What does a written "scope of work" often reveal?

3.4 If more than one estimator works on a project estimate (which is often the case), what is required for the success of the project?

3.5 a) What type of estimate organization could be used for a small plant?
b) What kind of organization is required for estimating a large complex installation?

3.6 How many codes of account should be used on each project? Why?

3.7 From the text, what does a designation such as 2V-101 reveal?

3.8 What is the primary purpose of a "code of accounts"?

3.9 In addition to costs, what should be determined as the estimate and equipment lists are being prepared?

3.10 What is required for project cost control?

3.11 The classification of the estimate is dependent upon what?

3.12 What determines the format of the capital estimate summary?

3.13 Why is a prime contractor desirable on a construction project?

3.14 In order to prepare a definitive estimate, what percentage of the project engineering should be completed?

3.15 Discuss the normal evolution of a project estimate.

3.16 Why is it important to organize and keep backup calculations and the final cost record?

4

Conceptual Estimating Using MasterFormat

No amount of good engineering will redeem a bad estimate.
Conversely, no 'good' estimate will redeem bad engineering.

INTRODUCTION

A cost estimator can be described as a cost engineer whose judgment and experience are utilized in the application of scientific principles and techniques to problems of cost estimating. There are as many paths that lead to a project cost estimate as there are cost estimators.

As is discussed in Chapter 2, a definitive cost estimate is expected to be accurate within a range of +15% and –5%.

"Definitive" and "accurate," in the same sentence as a +15% to –5% variance might appear self-contradictory, but history has demonstrated that

---

Contributed by William R. Barry, CCE, Scituate, Mass.

a 15–20% variance among the three lowest bidders on "hard money" cost estimates is not uncommon. Almost without exception, an estimator will swear that his or her estimate is "the right one," but the one estimate that will really be put to the ultimate test is the low bid. The contractor with the low bid must make good on the estimate, or the contract will suffer financial difficulties due to errors not corrected during the bidding period. The low bidder is sometimes described as the contractor who has not caught or corrected all the errors or omissions in the estimate—proof indeed that much truth is spoken in jest.

The focus of this chapter is the preliminary estimate for a project. The preliminary estimate is critical. The decision for a "go" or "no go" is in the balance. If a cost estimate is overestimated, there is a strong possibility that a good project may be scrapped because it is deemed too costly. Thus an opportunity is lost. On the other hand, if an estimate is too low, it is likely that the owner will overrun the schedule, the budget will overrun during construction, the owner will experience a longer payback period on the investment, added interest expense will be incurred during construction, etc. No amount of good engineering will redeem a bad estimate. Conversely no "good" estimate will redeem bad engineering, i.e., design faults or omissions which cause delays and cost overruns.

WORK BREAKDOWN STRUCTURES (WBS)

The WBS is the first task a cost estimator should address prior to assembling a cost estimate. This task should be a joint effort with the owner and the project design/construction team. A properly organized WBS will provide organization to the estimator for a systematic quantity takeoff. AACE International defines a WBS as:

> a product-oriented family tree division of hardware, software, facilities and other items which organizes, defines, and displays all the work to be performed in accomplishing the project objectives.

The WBS should be kept simple during the conceptual stages of a project estimate but be flexible enough to grow with the project from the budget through the final engineer's cost estimates. Contractors must assure that the WBS is in keeping with the project controls that the owner may require during construction. On many occasions the owner will carry the WBS forward to the owner's cost accounting system and into an operation and maintenance practice manual.

STANDARD CODE OF ACCOUNTS

Construction projects are designed and built by multidisciplined teams of experts representing diverse products and systems. Communication is very important between these experts, and the language must be understood by all. To facilitate this communication, MasterFormat has been established by the Construction Specifications Institute (CSI). Quoting CSI:

> MasterFormat serves the construction industry as a standard information filing and retrieval system and is an important tool for effective communication. It provides a uniform system for organizing information in project manuals, for organizing project cost data, and for filing product information and other technical data.*

Every estimator carries his or her own "book"; that is, every estimator has a specific way to assemble a cost estimate, either by system(s) or individual code of accounts. To avoid problems, a standard code of accounts must be established in an organized way. MasterFormat uses a 16–division format to classify codes of accounts and is commonly used by the majority of design professionals in the United States.

Each MasterFormat division is in turn broken down into a number of sections. For example, Division 11—Equipment has 35 sections, including number 11500—Industrial and Process Equipment, the primary area addressed in this book. The complete breakdown of MasterFormat divisions and section numbers and titles is shown in Table 4.1. The division and section numbers and titles are from *MasterFormat* (1988 edition). *MasterFormat* is published by the Construction Specifications Institute (CSI) and Construction Specifications Canada (CSC), and is used with permission of CSI.

Most computer-assisted estimating packages use the CSI MasterFormat to classify the work items in a cost estimate. The system is not perfect, but it is a good starting point. With some imagination, an estimator can adapt almost any project to follow in principle this system and to formulate a code

---

* *MasterFormat, Master List of Titles and Numbers for the Construction Industry*, Construction Specifications Institute, Alexandria, VA, 1988.

Table 4.1 MasterFormat Divisions and Sections

	Specifications
DIVISION 1—GENERAL REQUIREMENTS	
01010	SUMMARY OF WORK
01020	ALLOWANCES
01025	MEASUREMENT AND PAYMENT
01030	ALTERNATES/ALTERNATIVES
01035	MODIFICATION PROCEDURES
01040	COORDINATION
01050	FIELD ENGINEERING
01060	REGULATORY REQUIREMENTS
01070	IDENTIFICATION SYSTEMS
01090	REFERENCES
01100	SPECIAL PROJECT PROCEDURES
01200	PROJECT MEETINGS
01300	SUBMITTALS
01400	QUALITY CONTROL
01500	CONSTRUCTION FACILITIES AND TEMPORARY CONTROLS
01600	MATERIAL AND EQUIPMENT
01650	FACILITY STARTUP/COMMISSIONING
01700	CONTRACT CLOSEOUT
01800	MAINTENANCE
DIVISION 2—SITEWORK	
02010	SUBSURFACE INVESTIGATION
02050	DEMOLITION
02100	SITE PREPARATION
02140	DEWATERING
02150	SHORING AND UNDERPINNING

Table 4.1 (*Continued*)

02160	EXCAVATION SUPPORT SYSTEMS
02170	COFFERDAMS
02200	EARTHWORK
02300	TUNNELING
02350	PILES AND CAISSONS
02450	RAILROAD WORK
02480	MARINE WORK
02500	PAVING AND SURFACING
02600	UTILITY PIPING MATERIALS
02660	WATER DISTRIBUTION
02680	FUEL AND STEAM DISTRIBUTION
02700	SEWERAGE AND DRAINAGE
02760	RESTORATION OF UNDERGROUND PIPE
02770	PONDS AND RESERVOIRS
02780	POWER AND COMMUNICATIONS
02800	SITE IMPROVEMENTS
02900	LANDSCAPING
DIVISION 3—CONCRETE	
03100	CONCRETE FORMWORK
03200	CONCRETE REINFORCEMENT
03250	CONCRETE ACCESSORIES
03300	CAST-IN-PLACE CONCRETE
03370	CONCRETE CURING
03400	PRECAST CONCRETE
03500	CEMENTITIOUS DECKS AND TOPPINGS
03600	GROUT
03700	CONCRETE RESTORATION AND CLEANING
03800	MASS CONCRETE

Reproduced with permission from the Construction Specifications Institute, January 1995.

Table 4.1 (*Continued*)

DIVISION 4—MASONRY	
04100	MORTAR AND MASONRY GROUT
04150	MASONRY ACCESSORIES
04200	UNIT MASONRY
04400	STONE
04500	MASONRY RESTORATION AND CLEANING
04550	REFRACTORIES
04600	CORROSION RESISTANT MASONRY
04700	SIMULATED MASONRY
DIVISION 5—METALS	
05010	METAL MATERIALS
05030	METAL COATINGS
05050	METAL FASTENING
05100	STRUCTURAL METAL FRAMING
05200	METAL JOISTS
05300	METAL DECKING
05400	COLD FORMED METAL FRAMING
05500	METAL FABRICATIONS
05580	SHEET METAL FABRICATIONS
05700	ORNAMENTAL METAL
05800	EXPANSION CONTROL
05900	HYDRAULIC STRUCTURES
DIVISION 6—WOOD AND PLASTICS	
06050	FASTENERS AND ADHESIVES
06100	ROUGH CARPENTRY
06130	HEAVY TIMBER CONSTRUCTION
06150	WOOD AND METAL SYSTEMS
06170	PREFABRICATED STRUCTURAL WOOD
06200	FINISH CARPENTRY

Table 4.1 (*Continued*)

06300	WOOD TREATMENT
06400	ARCHITECTURAL WOODWORK
06500	STRUCTURAL PLASTICS
06600	PLASTIC FABRICATIONS
06650	SOLID POLYMER FABRICATIONS
DIVISION 7—THERMAL AND MOISTURE PROTECTION	
07100	WATERPROOFING
07150	DAMPPROOFING
07180	WATER REPELLENTS
07190	VAPOR RETARDERS
07195	AIR BARRIERS
07200	INSULATION
07240	EXTERIOR INSULATION AND FINISH SYSTEMS
07250	FIREPROOFING
07270	FIRESTOPPING
07300	SHINGLES AND ROOFING TILES
07400	MANUFACTURED ROOFING AND SIDING
07480	EXTERIOR WALL ASSEMBLIES
07500	MEMBRANE ROOFING
07570	TRAFFIC COATINGS
07600	FLASHING AND SHEET METAL
07700	ROOF SPECIALTIES AND ACCESSORIES
07800	SKYLIGHTS
07900	JOINT SEALERS
DIVISION 8—DOORS AND WINDOWS	
08100	METAL DOORS AND FRAMES
08200	WOOD AND PLASTIC DOORS
08250	DOOR OPENING ASSEMBLIES
08300	SPECIAL DOORS

Reproduced with permission from the Construction Specifications Institute, January 1995.

Table 4.1 (*Continued*)

08400	ENTRANCES AND STOREFRONTS
08500	METAL WINDOWS
08600	WOOD AND PLASTIC WINDOWS
08650	SPECIAL WINDOWS
08700	HARDWARE
08800	GLAZING
08900	GLAZED CURTAIN WALLS
DIVISION 9—FINISHES	
09100	METAL SUPPORT SYSTEMS
09200	LATH AND PLASTER
09250	GYPSUM BOARD
09300	TILE
09400	TERRAZZO
09450	STONE FACING
09500	ACOUSTICAL TREATMENT
09540	SPECIAL WALL SURFACES
09545	SPECIAL CEILING SURFACES
09550	WOOD FLOORING
09600	STONE FLOORING
09630	UNIT MASONRY FLOORING
09650	RESILIENT FLOORING
09680	CARPET
09700	SPECIAL FLOORING
09780	FLOOR TREATMENT
09800	SPECIAL COATINGS
09900	PAINTING
09950	WALL COVERINGS
DIVISION 10—SPECIALTIES	
10100	VISUAL DISPLAY BOARDS

Table 4.1 *(Continued)*

10150	COMPARTMENTS AND CUBICLES
10200	LOUVERS AND VENTS
10240	GRILLES AND SCREENS
10250	SERVICE WALL SYSTEMS
10260	WALL AND CORNER GUARDS
10270	ACCESS FLOORING
10290	PEST CONTROL
10300	FIREPLACES AND STOVES
10340	MANUFACTURED EXTERIOR SPECIALTIES
10350	FLAGPOLES
10400	IDENTIFYING DEVICES
10450	PEDESTRIAN CONTROL DEVICES
10500	LOCKERS
10520	FIRE PROTECTION SPECIALTIES
10530	PROTECTIVE COVERS
10550	POSTAL SPECIALTIES
10600	PARTITIONS
10650	OPERABLE PARTITIONS
10670	STORAGE SHELVING
10700	EXTERIOR PROTECTION DEVICES FOR OPENINGS
10750	TELEPHONE SPECIALTIES
10800	TOILET AND BATH ACCESSORIES
10880	SCALES
10900	WARDROBE AND CLOSET SPECIALTIES
DIVISION 11—EQUIPMENT	
11010	MAINTENANCE EQUIPMENT
11020	SECURITY AND VAULT EQUIPMENT
11030	TELLER AND SERVICE EQUIPMENT
11040	ECCLESIASTICAL EQUIPMENT

Reproduced with permission from the Construction Specifications Institute, January 1995.

Table 4.1 (*Continued*)

11050	LIBRARY EQUIPMENT
11060	THEATER AND STAGE EQUIPMENT
11070	INSTRUMENTAL EQUIPMENT
11080	REGISTRATION EQUIPMENT
11090	CHECKROOM EQUIPMENT
11100	MERCANTILE EQUIPMENT
11110	COMMERCIAL LAUNDRY AND DRY CLEANING EQUIPMENT
11120	VENDING EQUIPMENT
11130	AUDIO-VISUAL EQUIPMENT
11140	VEHICLE SERVICE EQUIPMENT
11150	PARKING CONTROL EQUIPMENT
11160	LOADING DOCK EQUIPMENT
11170	SOLID WASTE HANDLING EQUIPMENT
11190	DETENTION EQUIPMENT
11200	WATER SUPPLY AND TREATMENT EQUIPMENT
11280	HYDRAULIC GATES AND VALVES
11300	FLUID WASTE TREATMENT AND DISPOSAL EQUIPMENT
11400	FOOD SERVICE EQUIPMENT
11450	RESIDENTIAL EQUIPMENT
11460	UNIT KITCHENS
11470	DARKROOM EQUIPMENT
11480	ATHLETIC, RECREATIONAL, AND THERAPEUTIC EQUIPMENT
11500	INDUSTRIAL AND PROCESS EQUIPMENT
11600	LABORATORY EQUIPMENT
11650	PLANETARIUM EQUIPMENT
11660	OBSERVATORY EQUIPMENT

Reproduced with permission from the Construction Specifications Institute, January 1995.

Table 4.1 *(Continued)*

11680	OFFICE EQUIPMENT
11700	MEDICAL EQUIPMENT
11780	MORTUARY EQUIPMENT
11850	NAVIGATION EQUIPMENT
11870	AGRICULTURAL EQUIPMENT
DIVISION 12—FURNISHINGS	
12050	FABRICS
12100	ARTWORK
12300	MANUFACTURED CASEWORK
12500	WINDOW TREATMENT
12600	FURNITURE AND ACCESSORIES
12670	RUGS AND MATS
12700	MULTIPLE SEATING
12800	INTERIOR PLANTS AND PLANTERS
DIVISION 13—SPECIAL CONSTRUCTION	
13010	AIR SUPPORTED STRUCTURES
13020	INTEGRATED ASSEMBLIES
13030	SPECIAL PURPOSE ROOMS
13080	SOUND, VIBRATION, AND SEISMIC CONTROL
13090	RADIATION PROTECTION
13100	NUCLEAR REACTORS
13120	PRE-ENGINEERED STRUCTURES
13150	AQUATIC FACILITIES
13175	ICE RINKS
13180	SITE CONSTRUCTED INCINERATORS
13185	KENNELS AND ANIMAL SHELTERS
13200	LIQUID AND GAS STORAGE TANKS
13220	FILTER UNDERDRAINS AND MEDIA
13230	DIGESTER COVERS AND APPURTENANCES

Reproduced with permission from the Construction Specifications Institute, January 1995.

Table 4.1 *(Continued)*

13240	OXYGENATION SYSTEMS
13260	SLUDGE CONDITIONING SYSTEMS
13300	UTILITY CONTROL SYSTEMS
13400	INDUSTRIAL AND PROCESS CONTROL SYSTEMS
13500	RECORDING INSTRUMENTATION
13550	TRANSPORTATION CONTROL INSTRUMENTATION
13600	SOLAR ENERGY SYSTEMS
13700	WIND ENERGY SYSTEMS
13750	COGENERATION SYSTEMS
13800	BUILDING AUTOMATION SYSTEMS
13900	FIRE SUPPRESSION AND SUPERVISORY SYSTEMS
13950	SPECIAL SECURITY CONSTRUCTION
DIVISION 14—CONVEYING SYSTEMS	
14100	DUMBWAITERS
14200	ELEVATORS
14300	ESCALATORS AND MOVING WALKS
14400	LIFTS
14500	MATERIAL HANDLING SYSTEMS
14600	HOISTS AND CRANES
14700	TURNTABLES
14800	SCAFFOLDING
14900	TRANSPORTATION SYSTEMS
DIVISION 15—MECHANICAL	
15050	BASIC MECHANICAL MATERIALS AND METHODS
15260	MECHANICAL INSULATION
15300	FIRE PROTECTION
15400	PLUMBING
15500	HEATING, VENTILATING, AND AIR CONDITIONING
15500	HEAT GENERATION

Table 4.1 (*Continued*)

	15650	REFRIGERATION
	15750	HEAT TRANSFER
	15850	AIR HANDLING
	15880	AIR DISTRIBUTION
	15950	CONTROLS
	15990	TESTING, ADJUSTING, AND BALANCING
DIVISION 16—ELECTRICAL		
	16050	BASIC ELECTRICAL MATERIALS AND METHODS
	16200	POWER GENERATION—BUILT-UP SYSTEMS
	16300	MEDIUM VOLTAGE DISTRIBUTION
	16400	SERVICE AND DISTRIBUTION
	16500	LIGHTING
	16600	SPECIAL SYSTEMS
	16700	COMMUNICATIONS
	16850	ELECTRICAL RESISTANCE HEATING
	16900	CONTROLS
	16950	TESTING

Reproduced with permission from the Construction Specifications Institute, January 1995.

of accounts. There are exceptions to this rule. Some major industrial manufacturers have their own in-house code of accounts. They are usually tied into their operation and maintenance system. A CSI-formatted database can be crossreferenced and customized to fit their requirements.

This chapter presents cost factors to price project components such as sitework, yard piping, concrete slabs, walls, beams, masonry, structural and miscellaneous metal, moisture-proofing, doors, fenestration, finishes, process equipment, special construction, instrumentation/controls, conveying systems, piping and building services such as plumbing, HVAC, and electrical.

This chapter also applies the "Pareto Principle," or the "20/80 Rule." The Pareto Principle is based on the phenomenon of unequal distribution. Vilfredo Pareto, an economist, made extensive studies in the early 1900s about the unequal distribution of wealth. These studies have been applied to many fields, including construction cost estimating. The Pareto Principle suggests that 20% of the items in a conceptual estimate will represent 80% of the total costs. The balance of the items can be priced as a percentage the cost estimate. Most of the cost ratios or rules of thumb in the following text are guided by the Pareto Principle, concentrating on the CSI divisions which represent 80% of the total project cost for project facilities, i.e.,

Division 3—Concrete
Division 11—Equipment
Division 15—Mechanical
Division 16—Electrical

RULES OF THUMB

The following basic rules of thumb could be followed when pricing out a preliminary cost estimate. Like all data, these guidelines must be used with caution. The estimator "builds" an estimate; these rules of thumb do not.

Division 1—General Requirements

General requirements include supervision, field engineering, regulatory requirements, contract closeout, maintenance, distributables for the general contractor, etc. Also included here are *contingencies*:

A. When an estimate is assembled from a full set of documents, including drawings and specifications, use a 5% contingency. This contingency, *miscellaneous reserve* as some would like to call it, should be carried to the completion of the project.

B. If a reasonably good building layout, equipment list, site plan, and one-line process and electrical diagrams are available, use a 10% contingency.

C. If only a flow diagram is available and much is left to the estimator's discretion, use a 15% contingency.

Division 2—Sitework

Sitework includes site preparation, paving and surfacing, site improvements, drainage, landscaping, marine work, etc.

A. Assume that all building excavation work can be accomplished in an open cut, using 1:1 side slopes (or otherwise dictated by soil type).

B. Assume that dewatering should not be extraordinarily difficult. Use trenches/sumps or other dewatering systems as needed.

C. Assume that pile foundation or special work such as underpinning or steel sheeting are not required (unless otherwise indicated by soil studies or other historical data).

D. Sitework and yard piping will be priced at a percent of total project cost, usually 8–12% of project totals.

Division 3—Concrete

Concrete includes formwork, reinforcement, accessories, cast-in-place concrete, curing, grout, precast concrete, etc.

A. For a preliminary estimate, use the following minimum values (or as otherwise indicated by a specific design or requirement, e.g., a very high ground water level).
   1. Mats or grade slabs—12" minimum thick
   2. Exterior walls (structures)—12" minimum up to 15' high; over 15', 0.75" per foot (e.g., 18 × 0.75 = 13.5", say, 14").
   3. Interior walls (structures)—open both sides, 10" minimum up to 15' high; over 15', 0.65" per foot (e.g., 18 × 0.65 = 11.7", say, 12").
   4. Interior walls (structures) with water on one side—12" minimum up to 15' high; over 15', 0.75" per foot (e.g., 18 × 0.75 = 13.5", say, 14").
   5. Supported slab (building)—10" minimum.

6. Beams—average depth in inches = span (ft) × .75, e.g., 25' span × 0.75 = 18.75" or 18"; average width in inches = span (ft) × .5 (e.g., 25 span × 0.5 = 12.5", say, 12").
7. Columns—average width of beam squared.
8. Price of concrete using a composite unit cost to include accessories, forms, concrete, rebars, and finishes.
9. Reinforcing bars:
   a. Grade slabs 6" to 12" thickness at 100–125 lbs/cu. ft
   b. Elevated slabs 6" to 12" thickness at 125–150 lbs/cu. ft
   c. Beams/columns all sizes/thicknesses at 300–500 lbs/cu. ft
   d. Walls all heights/thicknesses at 150–200 lbs/cu. ft.
10. Average workhours for concrete complete, 4–7 wh/cu. yd
11. Average workhours for reinforcing steel, 10–12 wh/cu. yd
12. Average workhours for formwork, 0.10–0.40 wh/sq. ft
13. Average workhours for placing concrete, 0.50–1.0 wh/cu. ft

Division 4—Masonry

Masonry includes masonry, masonry grout, masonry accessories, unit masonry, stone, etc.

A. Exterior walls—use a composite price by system, i.e.:
   1. Brick, counter flashing, insulation, moisture-proofing, interior concrete masonry unit (CMU) 8" minimum, and coping (precast concrete, etc.).
   2. CMU (by style—split face, ground face, etc.), counter flashing, insulation, moisture-proofing, interior CMU 8" minimum, and coping (precast concrete, etc.).
   3. Stone, counter flashing, insulation, moisture-proofing, interior CMU 8" minimum, and stone coping.
   4. Reinforcing steel at masonry units—0.5 to 1.0 lb/sq. ft.

B. Interior walls—by type masonry block or structural facing tile:
   1. Masonry block—single wyth—4", 6", 8", 10", 12", etc.
   2. Structural facing tile—use double wyth block construction (single wyth walls are uneven). Use composite two-block combinations, e.g., 4" = 2" + 2" or 6" = 4" + 2" or 8" = 6" + 2", etc.

C. Take no deducts for openings unless they are major openings. Normally this amounts to 5–10% of total exterior surfaces, including windows and doors. By following this simple rule, you make allowances for lintels, special shapes, etc.

D. Fire-resistant rating:

Size	Description	Fire rating
4" lt wt	3–core/76% solid	1 hr
6" lt wt	3–core/53% solid	1 hr
8" lt wt	2–core/56% solid	2 hr
10" lt wt	3–core/55% solid	3 hr
12" lt wt	2–core/49% solid	4 hr

Division 5—Metals

Metals includes structural framing, joists, decking, fabrications, etc.

Structural metals—structural steel for buildings to include beams, columns, and bracing.

Water treatment facilities:

1. Average floor loading, 8 lb/sq. ft
2. Intermediate floor loadings, 10–12 lb/sq. ft
3. Heavy floor loadings, 15–25 lb/sq. ft
4. Structural frame/bar joint, 6 lb/sq. ft

Power plants:

1. Turbine rooms, 1.0–1.1 lb/cu. ft
2. Boilers, 1.3–1.5 lb/cu. ft
3. Precipitator, 1.1–1.3 lb/cu. ft
4. Buildings (support type), 0.6–0.7 lb/cu. ft
5. Platforms, 10–15 lb/sq. ft
6. Operating floors (turbines), 25–35 lb/sq. ft

Miscellaneous metals

Buildings/structures—miscellaneous metals include fabricated platforms, embedded iron/steel, metal bumpers, handrails, railings, ladders, grating/floor plates, lintels, and stairs.

Stairs—buildings having 100–5000 sq. ft/floor require one egress. Allow for one landing between floors. The width of a landing is equal to the length of the riser. The length of a landing is twice the length of the riser. Example: Height between basement and street/first floor = 20' and each riser measures 8" (0.67') by 4'. 20'/0.67 = 30 risers & 1 landing. Landing size = 4' × 8'.

Stairs—buildings having greater than 5000 sq. ft/floor require two egresses. Example: Plan area 60' × 150' = 9000 sq. ft/floor, having 20' as in the example above. 60 risers & 2 platforms.
Hatches/grating—allow 0.5% of floor space.

Miscellaneous metals, lintels, etc.—allow 1 lb/sq. ft of floor area:

1. Misc steel fabrication, ($Mat + 40%)
2. Misc galvanized steel fabrication, ($Mat + 30%)
3. Misc stainless steel fabrication, ($Mat + 18%)
4. Misc aluminum fabrication, ($Mat + 23%)
5. Handrails—steel painted, ($Mat + 16%)
6. Handrails—steel galvanized, ($Mat + 15%)
7. Handrails—stainless steel, ($Mat + 8%)
8. Handrails—aluminum, ($Mat + 14%)
9. Checkered plate—steel painted, ($Mat + 45%)
10. Checkered plate—steel galvanized, ($Mat + 40%)
11. Checkered plate—stainless steel, ($Mat + 30%)
12. Checkered plate—aluminum, ($Mat + 25%)
13. Stairs—steel painted, ($Mat + 22%)
14. Stairs—steel galvanized, ($Mat + 20%)
15. Stairs—steel metal pan, ($Mat + 22%)
16. Stairs—aluminum, ($Mat + 18%)
17. Grating—steel painted, ($Mat + 7%)
18. Grating—steel galvanized, ($Mat + 5%)
19. Grating—stainless steel, ($Mat + 2%)
20. Grating—aluminum, ($Mat + 4%)
21. Ladder—steel painted, ($Mat + 17%)
22. Ladder—steel galvanized, ($Mat + 15%)
23. Ladder—stainless steel, ($Mat + 11%)
24. Ladder—aluminum, ($Mat + 12%)

Division 6—Wood and Plastics

Carpentry includes rough carpentry and finish carpentry.

A. Temporary barricades at open excavations—allow 2 BFM/LF perimeter.
B. Roofing—allow 2 BFM/LF for cants and nailers at perimeter.
C. Temporary construction—allow 0.1 BFM/sq. ft of floor area.

Division 7—Thermal and Moisture Protection

Thermal and moisture protection—includes waterproofing, insulation, shingles, membrane roofing, flashing and sheet metal, skylights, metal siding and roofing, etc.

A. Roofing depending on engineer/architect preference, price the following composite systems:
   1. Built-up roof system—3" rigid insulation, 4–ply membrane, aluminum flashing, counter flashing (allow for 1.5' at perimeter), cap flashing.
   2. Engineered roofing systems—3" rigid insulation, engineered membrane, gravel, stepping pavers, counter flashing (allow for 1.5' at perimeter), cap flashing.

B. Moisture-proofing:
   1. Polyethylene membrane (10 mils) under floor slabs or occupied buildings.
   2. Bituminous black 40–50 mil thick at all exterior wall (with backer board where indicated) with earth backfill.
   3. Caulking and sealant—$0.50/sq. ft of building floor space (1995 cost basis).
   4. Flashing—could be included in wall systems.
   5. Counter flashing—exterior walls at perimeter (bottom or top of windows/door openings at masonry courses) or included with wall system.
   6. Brick/CMU exterior moisture-proofing (silicone-type spray coating).
   7. Hatches by type and metallurgy.

Division 8—Doors and Windows

Doors and windows includes metal doors and frames, wood doors and frames, entrances and storefronts, wood and metal windows, hardware, glazing, etc.

A. Doors—There are many varieties of doors. Keep this takeoff simple to include frame and hardware.
   1. Exterior/interior double doors, metal complete with hardware, closer, panic devices, etc.
   2. Exterior/interior single door, metal complete with hardware, closer, panic device, etc.
   3. Overhead doors complete with or without electric operator.

4. Exterior storefront type—aluminum including tempered glass and hardware, panic device, closer, etc.
5. Exterior truss-type door, steel (for crane opening, etc.) complete with hardware.
6. Special doors/hatches, e.g., watertight, complete with hardware, etc.
7. Interior by-fold doors complete with hardware.

B. Windows—There are many types and styles of windows. Takeoff should be a composite by metallurgy to include window and glass:
   1. Fixed window by metallurgy/composition, etc.
   2. Awning window by metallurgy/composition, etc.
   3. Window wall system (tube type) steel or aluminum frame.
   4. Composite/panels.
   5. When plans are not available, make an allowance of 5–10% (maximum) of all exterior surfaces of a building (occupied) for windows and glass. Check your building code.
   6. When a structure has only one means of egress, e.g., a stair building or small pump station, make no allowance for windows and glass (Interior electric lighting should be adequate).
   7. When a building is classified as a hazardous area, e.g., grit areas, wet wells, screening areas, make no allowances for windows and glass.

Division 9—Finishes

Finishes includes gypsum board, acoustical treatment, resilient flooring, carpet, special flooring, painting, wall coverings, etc.

Painting:
- Exterior walls (Use 1 times sq. ft of exterior walls)
- Interior walls (Use 2 times sq. ft of interior walls)
- Ceilings (Use sq. ft of building times number of ceilings)
- Floors (Use sq. ft of building times number of floors)
- Equipment (Use cubed surface area)
- Piping (1% of piping cost in Division 15)
- Sound attenuation (10% of wall area).

Floor finishes:
- Quarry tile (Add 25% to sq. ft area for cove bases)
- Ceramic tile (Add 25% to sq. ft area for cove bases)
- Terrazzo (Add 25% to sq. ft area for cove bases)

Resilient tile (Add 10% to sq. ft area for cove bases).
Whenever possible, try to price an item through a composite price, e.g., for floor finishes add 2.0' to the plan dimension to include a cove base.
Types of finishes:
Drywall—1–side, (composite stud/gypsum)
Drywall—2–sides, (composite stud/gypsum)
Plaster, (composite stud/lath)
Stucco, (3 coats/finish)
Fireproofing, (sprayed mineral fiber 1")
Intumescent epoxy fireproofing, 1–hr rating, (9 times mineral fiber cost)
Magnesium oxychloride fireproofing, 1–hr rating, (2 times mineral fiber cost)

Division 10—Specialties

Specialties includes flagpoles, identification devices, lockers, louvers, partitions, operable partitions, toilet accessories, etc.

Louvers:

A. Allow 10% of exterior walls for ventilation of combustion areas or incinerators.
B. Allow 20% of exterior walls for ventilation to provide combustion air for standby generators.
C. Allow 5% of exterior walls for ventilation of other areas.
D. Miscellaneous items—Allow 0.5% of building cost for maintenance and mechanical equipment areas.
E. Allow 1.0% of building cost for administration, labs, and personnel areas.

Division 11—Equipment

Equipment includes process equipment, pumps, and specialty equipment generic to the process. The following are rules of thumb for installation of steel fabricated equipment. This listing is based upon water and wastewater plants, but many items of equipment are commonly used in other processes and the rules of thumb apply for other applications. Additional equipment installation factors are given in Table 2.6.

Flow control at tanks or basins:
- Sluice gates, ($Mat + 20%)
- Slide gates, ($Mat + 25%)
- Bascule gates, ($Mat + 20%)
- Roller gates, ($Mat + 20%)
- Taintor gates, ($Mat + 20%)

Pumping equipment:
- Pumps—vertical/shafting, ($Mat + 25%)
- Pumps—vertical can/propeller/turbine, ($Mat + 20%)
- Pumps—vertical wet pit with rails, ($Mat + 25%)
- Pumps—vertical/horizontal—close coupled, ($Mat + 20%)
- Pumps—positive displacement—plunger, ($Mat + 20%)
- Pumps—positive displacement—screw, ($Mat + 20%)
- Pumps—chemical feed—plunger, ($Mat + 30%)
- Pumps—chemical feed—gear, ($Mat + 30%)
- Pumps—chemical feed—centrifugal, ($Mat + 30%)
- Pumps—sumps—duplex, ($Mat + 25%)
- Pumps—sumps—simplex, ($Mat + 30%)
- Pumps—sample type, ($Mat + 35%)
- Pumps—package pump stations, ($Mat + 20%)
- Sewage ejector—steel—simplex, ($Mat + 25%)
- Sewage ejector—cast iron—simplex, ($Mat + 20%)
- Sewage ejector—steel—duplex, ($Mat + 25%)
- Sewage ejector—cast iron—duplex, ($Mat + 20%)
- Hydro pneumatic systems, ($Mat + 30%)
- Anti-water hammer systems, ($Mat + 30%)
- Plant water system (LP/HP) skid mounted, ($Mat + 20%)
- Plant water system (LP/HP) pieces, ($Mat + 30%)

Screening and grit removal:
- Screens—mechanical type (all), ($Mat + 25%)
- Screens—manual (all), ($Mat + 20%)
- Grit—aerated tank/washer/classifier, ($Mat + 20%)
- Grit—chain/bucket/washer/classifier, ($Mat + 20%)
- Grit—detroiter/washer/classifier, ($Mat + 20%)
- Comminutors/barminutors, ($Mat + 20%)

Mixing systems:
- Mixers—turbine, ($Mat + 20%)
- Mixers—paddle type, ($Mat + 25%)
- Mixers—pump with rails type, ($Mat + 20%)
- Flocculators—paddle type, ($Mat + 20%)

Straight line sludge collectors are usually installed in a concrete tank/basin. They usually are installed in one or two pairs operated by a common motor/gear drive. Current U.S. practice allows stacked units (two deep). Currently in Japan there are plants with units up to three or four deep. Package plants, usually under 1.0 MGD, will have steel tanks included as part of the capital cost.

Clarifiers—straight line, ($Mat + 20%)
Clarifiers—x-collectors, ($Mat + 20%)
Clarifiers—straight line—1 tray, ($Mat + 20%)
Clarifiers—straight line—2 trays, ($Mat + 25%)

Circular or square clarifiers/concrete tanks with internal or external launderers. Package plants, under 1 MGD, will have steel tanks.

Clarifiers—reactor clarifiers, ($Mat + 30%)
Clarifiers—solids contact, ($Mat + 30%)
Clarifiers—upflow, ($Mat + 30%)
Clarifiers—primary, ($Mat + 20%)
Clarifiers—secondary without rapid sludge removal, ($Mat + 20%)
Clarifiers—secondary circular with rapid sludge removal, ($Mat + 25%)

Laminar sludge settling mechanisms are usually installed in a steel tank or as a retrofit to an existing process where limited space is available for expansion. Used mostly for thin sludges.

Clarifiers—laminar plate settlers, ($Mat + 25%)
Clarifiers—laminar tube settlers, ($Mat + 25%)

Trickling filters, shallow or deep bed, are usually installed in concrete tanks. Cost for the rock media is to be added to the shallow trickling filter, and cost for the plastic media or geometric shapes for biological growth must also be added to the deep bed units.

Trickling filter distributors, ($Mat + 20%)
Rotating biological contractors, ($Mat + 20%)

Aeration tanks/basins are either aerated by diffused air located at the outside/interior walls by either fixed or flexible air diffusers, a fixed grid of air diffusers fixed to the basins bottom slab, fixed mechanical aerators on concrete platforms or floating aerators with anchoring cables to the exterior basin walls.

Aeration—fine/coarse bubbles, ($Mat + 20%)
Aeration—channels with drop diffusers, ($Mat + 18%)
Aeration—channels with fixed diffusers, ($Mat + 20%)
Aeration—mechanical mixers—fixed, ($Mat + 20%)
Aeration—mechanical mixers—floating, ($Mat + 15%)

Chemical feed systems include the equipment to aid in the settling and/or conditioning of the sludges, adjusting the pH of the influent or effluent, disinfection of the effluent, etc.

Chemical feed systems—dry/wet, ($Mat + 30%)
Chemical feed systems—wet, ($Mat + 20%)
Silos (steel) storage—dry, ($Mat + 20%)
Bins (steel) bulk storage—dry, ($Mat + 20%)
Disinfection—chlorine systems, ($Mat + 25%)
Disinfection—ozone, ($Mat + 20%)
Grinders/comminutors, ($Mat + 20%)
Strainers—simplex/duplex, ($Mat + 20%)
Note: To calculate the total kW requirement, use the following guideline:
Sum of the total kW name plate times 3 for full voltage line starters.
Sum of the total kW name plate times 2 for equipment with reduced voltage starters that will be on standby duty.
Generators—without heat recovery, ($Mat + 18%)
Generators—with heat recovery, ($Mat + 30%)
Dewatering—vacuum filter, ($Mat + 25%)
Dewatering—belt press, ($Mat + 25%)
Dewatering—plate and frame press, ($Mat + 20%)
Dewatering—centrifuge, ($Mat + 20%)
Dewatering—hydro sieve, ($Mat + 20%)
Dewatering—belt press, ($Mat + 25%)
Thickening—dissolved air flotation with steel tank, ($Mat + 35%)
Thickening—dissolved air flotation with concrete tank—priced separately, ($Mat + 25%)
Note: Thickening equipment will have the same ratios as for the dewatering equipment listed above.
Scum concentrator with steel tank, ($Mat + 35%)
Scum concentrator with concrete tank by others, ($Mat + 25%)
Blowers—low pressure, centrifugal, ($Mat + 20%)
Blowers—low pressure, positive displacement, ($Mat + 20%)
Compressors—simplex/duplex, tank mounted, ($Mat + 25%)
Compressors—duplex, high pressure components, ($Mat + 35%)
Weirs/baffles and supports—all, ($Mat + 20%)
Troughs/weir and supports—all, ($Mat + 25%)

Maintenance equipment/shop tools as % of total plant cost:
Plant cost—less than $5 million, (0.75%)

Plant cost—$5 to $10 million, (0.50%)
Plant cost—$10 to $25 million, (0.25%)
Plant cost—$25 to $50 million, (0.20%)
Plant cost—$50 and greater, (0.15%)

Division 12—Furnishings

Furnishings includes casework and architectural specialties found in lavatories, offices, administration areas, etc.

Furnishings, etc., as % of total plant cost:
- Plant cost—less than $5 million, (0.20%)
- Plant cost—$5 to $10 million, (0.15%)
- Plant cost—$10 to $25 million, (0.10%)
- Plant cost—$25 to $50 million, (0.08%)
- Plant cost—$50 and greater, (0.05%)

Division 13—Special Construction

Special construction includes instrumentation. Electrical hookup/wiring is priced under Division 16—Electrical. Mechanical tubing for pneumatic control systems should be included with this division. All ratios to material cost include testing and startup.

Instrumentation and controls (I&C)—% of total plant cost:
- I&C—not heavily computerized, (3%)
- I&C—heavily dependent on computerization, (5%)

Chemical storage tanks:
- Tanks—fiberglass/PVC, ($Mat + 25%)
- Tanks—polyethylene/PVC, ($Mat + 25%)
- Tanks—steel, ($Mat + 30%)
- Tanks—steel rubber lined, ($Mat + 20%)
- Note: Above data is for tanks with capacities 500–5000 gal.

Water reservoirs/tanks/towers:
- Water reservoirs/tanks/towers are usually priced as a subcontracted item by the prestressed concrete or steel tank designer/fabricator/installer. Add 5% to subcontractor's quote for general contractor's overhead and profit.
- Caution: The sub-bidder on the steel tanks usually subcontracts the foundation/ring wall. Add for this cost in addition to the steel tank. The foundation is normally included with the prestressed

concrete tank sub-bid. The sitework and yard piping is usually not included with either sub-bid.

Digester equipment:

Sludge digester—covers, ($Mat + 30%)

Sludge digester—heat exchanger, ($Mat + 25%)

Sludge digester—gas mixing, ($Mat + 35%)

Sludge digester—gas equipment, ($Mat + 35%)

Special equipment—filtration:

Filter bottoms—plastic cellular type, ($Mat + 30%)

Filter bottoms—pipe type, ($Mat + 25%)

Filter media—activated carbon, ($Mat + 18%)

Filter media—coal/anthracite, ($Mat + 20%)

Filter media—gravel(s)/sand, ($Mat + 25%)

Pressure filter system, complete, ($Mat + 25%)

Automatic backwash filter, complete, ($Mat + 35%)

Note: Concrete or steel structure is not included in the filter pricing above.

Division 14—Conveying Systems

Conveying systems includes conveyors, elevators, hoisting equipment, bridge cranes, monorails, etc.

Conveyors—straight/inclined, ($Mat + 20%)

Conveyors—elevated straight/inclined, ($Mat + 25%)

Conveyors—screw type—carbon steel, ($Mat + 20%)

Conveyors—screw type—stainless steel, ($Mat + 15%)

Bridge cranes, ($Mat + 20%)

Gantry cranes, ($Mat + 20%)

Monorails/hoist, ($Mat + 25%)

Division 15—Mechanical

Mechanical includes all process piping systems including pipe, fittings and valves, plumbing, fire protection, heating/ventilation/air conditioning, etc. The mechanical piping system cost ratios stated below are based on the installed cost of the equipment described and listed in Divisions 11 and 13.

Pumping systems:

Pump—0.1 to 1 MGD-CS, ($M+L × 100%)

Pump—1 to 5 MGD-CS, ($M+L × 100%)

Pump—5 to 10 MGD-CS, ($M+L × 80%)

- Pump—10 to 25 MGD-CS, ($M+L × 50%)
- Pump—25 to 50 MGD-CS, ($M+L × 40%)
- Pump—50 to 100 MGD-CS, ($M+L × 35%)
- Pump—0.1 to 1 MGD-VS, ($M+L × 60%)
- Pump—1 to 5 MGD-VS, ($M+L × 50%)
- Pump—5 to 10 MGD-VS, ($M+L × 50%)
- Pump—10 to 25 MGD-VS, ($M+L × 40%)
- Pump—25 to 50 MGD-VS, ($M+L × 30%)
- Pump—50 to 100 MGD-VS, ($M+L × 25%)

Miscellaneous pumping systems:

- Hydro pneumatic system, ($M+L × 45%)
- Anti-water hammer system, ($M+L × 25%)
- Plant water system, ($M+L × 75%)
- Grit system (pumped/air lift), ($M+L × 45%)
- Grit system (chain/bucket), ($M+L × 25%)
- Ejector with pot—simplex/duplex, ($M+L × 35%)
- Chemical feed systems, ($M+L × 25%)
- Disinfection (chloriation/ozone), ($M+L × 25%)
- Grinders (floor mounted/in-line), ($M+L × 45%)
- Tanks—chemical storage, ($M+L × 45%)

Aeration systems:

- Blowers—400 to 2000 CFM-CS, ($M+L × 90%)
- Blowers—2000 to 5000 CFM-CS, ($M+L × 75%)
- Blowers—5000 to 10,000 CFM-CS, ($M+L × 60%)
- Blowers—10,000 to 20,000 CFM-CS, ($M+L × 50%)
- Blowers—20,000 to 30,000 CFM-CS, ($M+L × 30%)
- Blowers—400 to 2000 CFM-VS, ($M+L × 55%)
- Blowers—2000 to 5000 CFM-VS, ($M+L × 45%)
- Blowers—5000 to 10,000 CFM-VS, ($M+L × 35%)
- Blowers—10,000 to 20,000 CFM-VS, ($M+L × 30%)
- Blowers—20,000 to 30,000 CFM-VS, ($M+L × 20%)
- Channel/tank diffuser fixed/swing, ($M+L × 100%)
- Air stripping horizontal/vertical—vessels, ($M+L × 35%)
- Carbon absorption horizontal/vertical—vessels, ($M+L × 35%)

Dewatering/thickening:

- Screens-mechanical, ($M+L × 15%)
- Vacuum filter, ($M+L × 35%)
- Belt press, ($M+L × 15%)
- Plate and frame press, ($M+L × 10%)
- Centrifuge, ($M+L × 15%)

Hydro sieve, ($M+L × 15%)
Dissolved air flotation, ($M+L × 45%)
Scum concentrator, ($M+L × 30%)

Compressed air systems:
Compressor and tank—package, ($M+L × 100%)
Compressor—plant air, ($M+L × 100%)

Electrical power generation:
Generator—stand-by power, ($M+L × 15%)
Generator—prime power, ($M+L × 35%)
Generator/cogeneration, ($M+L × 60%)

Sludge digestion:
Heat exchanger, ($M+L × 45%)
Boiler/heat exchanger, ($M+L × 45%)
Gas mixing equipment, ($M+L × 65%)
Gas compressor, ($M+L × 100%)

Piping system:

The ratio of process pipe systems cost to pipe/fittings/valve cost is a factor of 2.5. For example, if the pipe has a value of 1, the total price of the installed piping system would be 2.5 times the cost of the pipe. Insulation, thermal or otherwise, can be found on process equipment and piping where there would be contact with plant personnel and/or to maintain an even temperature in a facility. For fittings, allow 3 times pipe length measured along the fitting centerline.

Plumbing:

Plumbing, if priced separately, should be priced by assembly. The average cost per assembly, installed to include the fixture, all rough-in piping, supports, cold/hot water supply piping and waste/vent piping is 5 times the materials cost of the fixture. This would include the labor and all accessories to install the assemblies.

Sump pumps should be priced per assembly. The average cost per assembly, installed to include all piping, valves and supports is 3 times the materials cost of the sump pump.

Fire protection has an average cost of 1–2% of the cost of the building where required.

Heating, ventilation and air conditioning (HVAC):

HVAC has an average cost of 6–8% of a building such as an administration building and includes administrative area, laboratories, shower/toilet facilities, storage area, etc. HVAC, if priced separately, should be priced by assembly. The average cost per assembly, installed to include the fixture, all rough-in piping, supports, hot water supply and return piping and valves, is 10 times the materials cost of the fixture. This would include the labor and all accessories to install the assemblies.

Division 16—Electrical

Electrical systems include basic electrical bulk materials, power generation, low/medium/high voltage distribution, service distribution, special systems, communications, electric resistance heating, control wiring and testing.

NEMA—Code definitions:

NEMA Type 1—general purpose enclosure.

NEMA Type 4 and 4X—watertight and dust-tight enclosure, stainless steel.

NEMA Type 4XO—watertight, dust-tight and corrosion resistant, polyester enclosure.

NEMA Type 7 and 9O—hazardous locations, Class I, Groups C and D; Class II, Groups E, F, and G.

NEMA Type 12—dust-tight and drip-tight industrial use enclosure.

Power output devices: Cost differential from base—NEMA Type 1 rated at 600 volts maximum.

Size	NEMA 1	NEMA 4 & 4X	NEMA 4XO	NEMA 7 & 9O	NEMA 12
0	1.0	1.6	1.9	2.0	1.15
1	1.0	1.6	1.8	1.8	1.15
2	1.0	1.7	1.9	1.8	1.15
3	1.0	1.7	1.9	2.1	1.10
4	1.0	1.4	1.5	1.8	1.10
5	1.0	1.5		–	1.10
6	1.0	1.1		–	1.05

Motor connection assemblies includes the cost of motor starter, breaker, conduit, wire, terminations, disconnect switch, and flexible connector priced as an assembly cost.

Motor connection assemblies—460 volts

15 hp and smaller:
- Motor starter—Size 1
- Breaker—30 amps
- Conduit—100 LF at 0.75"
- Wire—400 LF 1/C #10
- Terminations 16 ea #10
- Disconnect switch—30 amps
- Flexible motor connector

20–30 hp:
- Motor starter—Size 2
- Breaker—60 amps
- Conduit—100 LF at 0.75"
- Wire—400 LF 1/C #8
- Terminations 16 ea #8
- Disconnect switch—60 amps
- Flexible motor connector

40–60 hp:
- Motor starter—Size 3
- Breaker—100 amps
- Conduit—100 LF at 1.0"
- Wire—400 LF 1/C #4
- Terminations 16 ea #4
- Disconnect switch—100 amps
- Flexible motor connector

75–125 hp:
- Motor starter—Size 4
- Breaker—200 amps
- Conduit—100 LF at 1.50"
- Wire—400 LF 1/C #2/0
- Terminations 16 ea #2/0
- Disconnect switch—200 amps
- Flexible motor connector

150–250 hp:
- Motor starter—Size 5
- Breaker—400 amps
- Conduit—100 LF at 2.0"
- Wire—400 LF 1/C #300 MCM
- Terminations 16 ea #300 MCM
- Disconnect switch—400 amps
- Flexible motor connector

300–400 hp:
- Motor starter—Size 6
- Breaker—600 amps
- Conduit—100 LF at 4.0"
- Wire—400 LF 1/C #500 MCM
- Terminations 16 ea #500 MCM
- Disconnect switch—600 amps
- Flexible motor connector

CONCLUSION

Conceptual estimating using rules of thumb or cost factors works reasonably well when the factors are used in their proper context. These factors are by no means the only ones to follow when preparing a preliminary cost estimate. They have been developed over the past 20–some years. A cost estimator's education is cumulative. It is an assembly of historical (or hysterical) data based on one's professional experience. To quote Nathan Gutman:

> . . . the nature of the experience curve is strictly phenomenological, that is, it is an observable phenomenon not related to any laws of the universe. One of the pitfalls is trying to understand why this phenomenon works and trying to make it fit existing laws. Trying to find an exact explanation of the experience curve can be a futile and exasperating exercise, because of the complexity and multitude of human and machine variables that need to be correctly aligned.*

Conceptual cost estimates fall into the same category as an experience curve; they are based on one's professional experience over the years. In addition, a good conceptual cost estimate is a balanced combination of science, art, years of experience and occasional diet of humble pie, for there are no absolutes, just estimates.

NOMENCLATURE

BFM board foot measure

---

* Nathan Gutman, *How to Keep Product Costs in Line*, Marcel Dekker, New York, 1985.

cu. ft	cubic foot
CFM	cubic feet per minute
CMU	concrete masonry unit
CS	constant speed
CSI	Construction Specifications Institute
cu. yd	cubic yard
hp	horsepower
HP	high pressure
HVAC	heating, ventilating, and air conditioning
I&C	instrumentation and controls
kW	kilowatt
LF	linear foot
LP	low pressure
lt wt	light weight
$Mat	Material or equipment cost
$M+L	installed cost of equipment item
MGD	million gallons per day
NEMA	National Electrical Manufacturers Association
PVC	polyvinyl chloride
sq. ft	square foot
VS	variable speed
wh	workhours

QUESTIONS

4.1 What is the most common system used by design professionals to classify codes of accounts?

4.2 What four CSI cost divisions account for 80% of total project cost for most process facilities?
4.3 What is the name of the concept that 20% of the items in a project will account for 80% of the cost?
4.4 Brick and block are included under which CSI division?
4.5 Piping is included under which CSI division?
4.6 What contingency allowance is suggested if the project is conceptual and only a flow diagram is available?
4.7 List some of the items which are considered to be sitework.
4.8 What is the suggested minimum thickness for a mat or grade slab?
4.9 What is the fire resistance rating of a 3–core, 55% solid, 10–inch light weight masonry wall?
4.10 What is the installation labor factor for steel galvanized handrails?
4.11 How much wood should be allowed for temporary barricades at open excavations?
4.12 When plans are not available, what allowance should be made for window and glass area in the estimate for occupied building?
4.13 When estimating the cost of tanks to be installed by a subcontractor, how much should be added to the subcontractor's quote to allow for the general contractor's overhead and profit?
4.14 a) What percentage of administrative building costs is attributable to plumbing? b) to fire protection? c) to HVAC?

5

Retrofitting and Expansion of Existing Plants

INTRODUCTION

Many years of experience in performing various estimates dictated by the circumstances at the time suggests that retrofit estimating is the most difficult of all situations for the estimators. Today, to avoid the expense of building new plants, retrofitting is becoming a major part of capital cost determination for many companies in the process industries. Huge capital outlays are also being spent for environmental and safety considerations.

A myriad of problems not encountered in new construction projects are generated during work in an *operating* plant. These include, among other considerations:

1. restricted/limited accessibility
2. operating plant safety considerations
3. required electric power source and motor control center revisions
4. outdated as-built drawings
5. pipe-rack loading limitations

6. control systems revisions and/or limitations
7. plant air, steam, and water revisions and/or limitations
8. complicated civil scope
9. complicated piping and mechanical scope

From the above list, it is apparent that a detailed "scope of work" must be prepared if one is to make an acceptable capital estimate of a retrofit project.

There are no established standards for retrofit work; each project must be evaluated on the unique circumstances of the job. Historical data from similar work in a similar situation is of great value. However, application of new plant standards to retrofit work would be catastrophic. The estimate would inevitably be low if new plant standards were applied. It is always more difficult and more expensive to work in an operating plant.

ACCESSIBILITY

Access to the work area in an operating plant is almost always a problem but varies considerably within a plant. For example, a remote tank farm affords reasonable access while a congested area affords very limited access. The work must be scheduled to take full advantage of off-hours, slow times, etc., so as *not* to impede the work unnecessarily. In addition, in any operating plant, all construction work stops during safety alerts, and this slows work in all areas of a plant. On one job for a major chemical company it was necessary to replace a heat exchanger that the plant had completely surrounded. It was necessary to bring in a "cherry picker" to remove the machine up through the structure and out the top. The cost of removing the machine was much greater than the cost and installation of the new exchanger.

In almost all operating chemical plants, operations take precedence over construction, so access is slowed by operation of nearby units, plant traffic, etc. Accessibility is a major consideration on retrofit projects.

PLANT SAFETY CONSIDERATIONS

Safety considerations impinge upon all construction work in an operating plant. Safety requirements are strictly enforced in all chemical and petrochemical plants. Examples of safety considerations include welding bans in some areas of almost all plants and stringent electrical codes, often requiring plant safety engineering approval, for work. All plants require safety equipment such as emergency air packs in certain areas, and personnel, including construction workers, must be trained to use such equipment. All plants require safety orientation for construction personnel.

Keep in mind that construction work stops during a safety alert. Other considerations include requirements of the Occupational Safety and Health Administration (OSHA). It seems that someone from the federal government is in any large chemical installation every day for some reason or other. These people often dictate how work can be done.

ELECTRICAL POWER SOURCE AND MOTOR CONTROL CENTERS

Electrical work in most chemical plants is unbelievably slow, so that it is imperative that the "scope of work" for the electrical portion of all retrofit projects be as detailed as possible. The scope must include conduits (ft), wire (ft), fixtures, switches, etc. If power-driven equipment is used, the motor control center must be located and motor starters and all peripheral requirements identified. Almost all plants have unique electrical requirements; thus it is important that the engineer doing the scope is familiar with any unique provisions of the plant where the work is to be done.

Always on a retrofit project, glean as much information as possible from the plant operating and maintenance personnel concerning the unique requirements of the plant.

OUTDATED AS-BUILT DRAWINGS

A major problem, particularly in plants more than ten years old, is drawings that have *not* been updated to reflect what is actually in the field. For example, excavations for pipelines, drains, foundations, etc., are particularly affected when a line is unearthed that does not appear on the drawings. This causes delays and sometimes abandonment of the original project scope. It is also sometimes impossible to reconcile from file drawings what is in the field. Although correct drawings are best, one can see what is above ground and make necessary scope changes as required. What is below the ground is another matter.

Remember, in retrofit estimates, one *cannot* assume that as-built drawings are correct.

PIPE-RACK LOADING

It is most important, if new pipelines are to be added, to ascertain if the existing pipe-racks will be adequate support for the new line or lines. Considerable expense can be incurred to supply new supports or to remove dead lines from existing racks to make room for the new lines.

In old plants many lines may be inoperative and may not have been operating for several years. Therefore one cannot assume that any line is really available until the status of the line has been checked.

On a small project, 500 feet of 2" pipe not anticipated can ruin a capital estimate.

CONTROL SYSTEMS

Experience shows that this is the account that is most often underestimated on a retrofit project. It is essential that experienced control systems engineers scope the work and itemize the required components for the system.

Counting the control loops and allowing a standard cost for each loop is absolutely inadequate for reasonable estimating accuracy on a retrofit project. Control loop costs vary from a few hundred dollars to many thousands of dollars; therefore, a standard cost does *not* exist.

All chemical and petrochemical plants have established methodologies for installing instruments. It is most important to consider the proscribed methods of installation at the particular location of the project and to account for all the bulks, such as wire, conduit, tubing, etc., which are required.

In many retrofit projects, the existing control board is not large enough to accommodate additional instruments, alarms, etc. If this is the case, a new board must be furnished. One must exercise extreme care when estimating control systems on a retrofit project.

AVAILABILITY OF PLANT AIR, STEAM, AND WATER

Within the confines of an existing operating plant, things are not always as they appear to be. For example, a steam line may be nearby but not have adequate additional capacity for the project under consideration. This is also true of air and water lines. Thus, it is necessary to check with plant personnel before adding anything to an existing line.

On most retrofit projects, changes of scope are difficult after the work is underway and significant additions to the scope on small jobs will cause the estimate to exceed all allowances for overrun. It suffices to state that a source of steam, air, and water must be positively identified before work is started.

CIVIL ENGINEERING SCOPE

One of the prevalent problems concerning civil scope is the excavations. As mentioned earlier, unexpected buried pipelines can destroy the scope as written. The other major problem in existing plants is whether or not the excavated soil is contaminated. The cost of disposal of excess contaminated soil can be 25–30 times as great as uncontaminated soil disposal. Thus the determination must be made early in the project.

All the work, such as structural steel requirements (itemized), concrete details, etc., must be carefully scoped. In many retrofit projects civil engineering is the major account.

PIPING AND MECHANICAL SCOPE

For retrofit estimating, this may be the account most difficult to scope. This scope includes pipe, valves, materials of construction, etc. A complete descriptive valve list is an absolute necessity for an accurate estimate. For example, a 2" valve can be purchased for as little as $30 or cost several thousands of dollars, depending on the service. The same wide range of costs for pipe also exists. That is, 2" pipe can vary from less than $2/foot to over $50/foot for high alloy lined pipe. Therefore, a complete descriptive scope is required.

OTHER CONSIDERATIONS

If the retrofit project is in a computer controlled area, the effect on the computer system must be ascertained and any subsequent modification to the existing system considered in project evaluation. Failure to consider the above can be catastrophic on a small project.

For any significant retrofit project P&IDs are required. The work required to produce the drawings eliminates omissions and assures that the P&ID systems are properly designed.

Keep in mind that in an operating plant, operations almost always take preference over construction. Therefore, unforeseen delays will occur on most projects, requiring some allowance by the estimator. There is no set percentage for such an allowance. Experience shows the only way to estimate such a contingency is from careful study of the actual job, conferring with operating and plant personnel, and studying the operating logs of nearby processing units.

Little mention has been made of equipment cost determination in this discussion because equipment cost can be determined very accurately from vendors, historical records, etc. However, the specifications for equipment must be well written to eliminate misunderstanding between the vendor and the buyer.

Remember that the estimate can never be better than the scope. The scope must always be as detailed and complete as possible.

QUESTIONS

5.1 Why is it dangerous to use historical data from new plant construction to estimate retrofit work?

5.2 In retrofit work, what takes preference: plant operations or construction activities?

5.3 What is the effect on construction work when a safety alert occurs in an operating plant?

5.4 In retrofit work, should the contractor rely on as-built drawings of the plant? Why or why not?

5.5 In retrofit work, what problems are common in using existing pipe-racks to support new lines?

5.6 What cost account is most often underestimated on retrofit projects?

5.7 What problem commonly occurs in retrofit work with plant air, steam, and water lines?

5.8 Name two of the most prevalent problems concerning civil scope in retrofit projects.

5.9 What is the single most important prerequisite to preparing an acceptable capital estimate for a retrofit project?

6

Labor Estimates

Of all the elements that make up a capital estimate for a construction project, labor is the most illusive and affords much latitude for error. The final labor component is known for certain only after all of the bills for the project have been paid, if complete records have been kept throughout the project life. Unfortunately, many project staffs keep insufficient records to substantiate a detailed breakdown of the costs even at project completion.

Labor is difficult to estimate because of the differences in productivity in various geographic areas, whether or not it is a union job, labor and craft availability at the plant site and many other semisocial considerations such as opening day of hunting season. If the economic conditions are somewhat depressed, then labor is available and turnover is low. However, if the economy is on an upswing, labor is difficult to recruit and turnover is high.

From the above, it is obvious that a very careful analysis needs to be completed which is specific to the project location and considers such things as the mores and traditions of the workforce, the general economic conditions, the local union scale (if a union job), the prevailing wage rate (non-union), productivity numbers applicable to the area, etc. Keep in mind that labor will constitute 25–40 percent of the total project cost and is very difficult to control and/or estimate.

In special cases, such as remote areas, housing may need to be furnished and other labor-related expenses considered, for example, catering services, etc. One must also consider local infrastructure, such as roads and schools, to plan adequately for a large influx of construction workers into a sparsely populated area.

ESTIMATES OF LABOR FOR AN ORDER-OF-MAGNITUDE ESTIMATE

A rough order-of-magnitude estimate of labor is sometimes made for the most preliminary of estimates. Percentages may be used to proportion an order-of-magnitude estimate into the various major divisions of a capital cost estimate such as: bare equipment costs, bulk accounts, labor, etc. However, the accuracy achieved is less than the anticipated accuracy for the total estimate and thus labor estimating is not a recommended practice. The accuracy for the overall estimate is at best –30 to +50% (see Chapter 2). Thus, an estimate of labor based on the above is little more than an educated guess.

ESTIMATES OF LABOR FOR A FACTORED ESTIMATE

The estimating procedure proposed in Chapter 2 for factored estimates provides at least a reasonable estimate of the total labor required for a project. Although the labor hours are not determined as such, a dollar amount for the labor is estimated. Then by estimating the weighted average wage rate for the various tasks, the dollar amounts may be expressed in workhours. For example, refer to the equipment and installation cost summary for the ABC Alcohol Company's fermentation plant shown in Table 2.8. In this table the labor column expresses the labor in dollars required for setting the equipment and for the bulk accounts. Then using the procedure described above, an estimate of the workhours may be readily computed.

An estimate of the labor required for the field is described above, combined with about 35% of total field indirects, allows one to compute an estimate of total field workhours. A similar methodology may be used to estimate the engineering and home office hours from the dollar allowances on the form. Thus, a total workhour estimate may be made of the project by combining the estimates for all the plants and/or cost centers. Thus, workhour estimates can be made relatively early in a project and refined as the project develops, much like the capital estimate.

The workhours to set equipment are usually taken from a historical data base and most engineering construction contractors have a systematic method for assigning such hours, such as a book or listing for different types of equipment.

The bulk account workhours may be calculated using either workhours per unit (e.g., workhours per yard of concrete) or workhours per thousand dollars of bulk material costs. These methods require the estimator to make judgments based on experience and an analysis of the account. The labor required to install concrete depends greatly upon whether it is a large block foundation, a wall, a floor, a pad, a footer, etc. Therefore, it is necessary to examine the account to assign a reasonable estimate of workhours. This is also true of the other bulk accounts. That is, an analysis must be made to assign realistic workhour allowances.

Workhour projections made on the basis of a purely factored estimate may be 35% too low to 15% too high but still are much better than no estimate at all.

ESTIMATE OF LABOR FOR A PRELIMINARY (BUDGET AUTHORIZATION) ESTIMATE

The labor component becomes considerably more accurate for a preliminary estimate because of the additional engineering which has been completed on the project. That is, the process flowsheets are complete, the bulk accounts better defined, the utility requirements known, the control loops counted, etc., giving a much improved basis for the estimation of workhours.

Also, at this stage, at least a preliminary (milestone) schedule has been developed which gives a more realistic view of the labor loading and allows the initiation of labor leveling. Good planning and scheduling is an absolute must for proper utilization of construction labor. That is, unless great care is exercised in planning the work, the labor will not be utilized to get even average productivity from a construction labor force.

At the preliminary estimate stage, the labor required to set equipment is more nearly correct because at least cursory specifications and data sheets have been prepared to allow telephone solicitation of costs for the major equipment items.

The bulk accounts for a preliminary estimate are also improved over those for a factored estimate. Either the estimate of the factors is improved or a preliminary takeoff has been made which allows a considerably improved estimate of workhours for each account. The bulk accounts which are frequently underestimated for a processing plant are piping, electrical (including instrumentation), and concrete. Therefore, the aforementioned accounts need to be defined well as early as possible in the evolution of a project.

Workhours for setting equipment will at times be estimated by the vendor. If this is the case, be sure that the vendor and the contractor are estimating the same package of work. Vendor estimates tend to be low while contractor estimates tend to be high. Do not get into a confrontational situation but rather promote understanding between the parties for a more harmonious working situation and a reasonable estimate of the actual workhours required.

Engineering and home office workhours may be better estimated because of advances in engineering and planning for the project. Since process flowsheets and other engineering documents are available, an accurate estimate of engineering and home office hours should be possible at this time in the evolution of a project.

DEFINITIVE ESTIMATE WORKHOURS

The engineering is about 40–45% complete with 90–95% of the major equipment items quoted at the time the definitive estimate is made in the normal evolution of a construction project. Thus the basis for workhour estimates is much improved over the preliminary (budget authorization) estimate discussed previously.

The schedules are developed for the individual plants and cost centers to a much greater extent than before. The critical path model (CPM) should be nearing completion, allowing more definition of workhour requirements.

DETAILED ESTIMATE WORKHOURS

Workhour requirements for the detailed estimate are based on actual labor-loaded schedules, work breakdown structures, actual hours to set equipment, and detailed breakdown of bulk accounts and subsequent workhour estimates. However, workhours are always estimates and afford the most frequent source of error in a detailed estimate.

One of the most prevalent problems near the completion of a project is the labor productivity slowdown which occurs almost universally. The slowdown occurs whether or not crafts and laborers are readily available or in short supply. If labor is plentiful, the concern becomes one of employment after the present job is finished. If labor is in short supply, the workers get other jobs before the project is finished. In either situation, labor productivity suffers.

Estimators should be aware of the problem and take measures to mitigate the effect of same, such as detailed planning. However, an allowance in the estimate to account for the inevitable is wise.

WORK BREAKDOWN STRUCTURE AND LABOR PLANNING

One of the most difficult tasks on any construction project is providing sufficient planning to allow a breakdown of the work into small day-to-day work packages that facilitate the proper utilization of construction crews. Nothing is as detrimental to the morale of workers as being on the job without the work being planned well enough for them to be kept busy. The lack of such planning is absolutely the fault of the construction management.

The United States federal government refers to such planning as the work breakdown structure (WBS). However, labor planning seems to be a better description of actual needs. The job must be planned with an overall view of the craft and labor requirements and then the work coordinated in a way which allows craft leveling with as full a utilization of people as is possible.

Labor planning requires detailed scheduling with time constraints and labor loading for each day of the project duration. This allows an orderly buildup of crafts and labor as well as an orderly buildup of the capital cost estimate.

For megaprojects (say more than $1 billion), the detailed scheduling required is a time-consuming task, but the time is well spent and actually results in cost savings through efficient use of resources and the leveling of craft and labor requirements.

Craft and labor leveling simply entails taking the total contingent of people on the job at any particular time during the construction of the project and making sure that demand and supply match. For example, the work for the pipefitters must be planned to allow an orderly buildup, utilization at peak demand, and then an orderly reduction of the craft force at the end of the project.

Scheduled overtime is counterproductive and can result in substantially increased costs with no net gain in productivity. In fact, sustained scheduled overtime can result in a net loss of output as compared to the work which would have been accomplished on a regular nonovertime schedule. The detrimental effects of scheduled overtime are discussed more fully in Chapter 10.

Labor planning requires that the workers needed for each task be estimated well in advance. It follows that some estimates will be high and others low. Therefore, construction management must maintain sufficient flexibility to make minor corrections as the job proceeds, sometimes day to day. For example, work may be delayed on some part of the job for a multitude of reasons. If this occurs, a contingency plan for the use of the crafts people and laborers somewhere else on the job is a necessity. Similarly, when work is delayed, overtime should be avoided except for emergency situations of short duration.

Coordination among frontline supervisors, engineers, construction experts, and construction management is a must for well-executed construction projects.

QUESTIONS

6.1 Why should complete capital cost records be kept for a construction project?

6.2 What are the effects of the general economic condition on recruiting and maintaining a construction workforce?

6.3 What are the special considerations necessary for recruiting labor in a remote location?

6.4 What is the conclusion drawn for labor estimates on an "order-of-magnitude" estimate?

6.5 In a factored estimate, how is labor expressed?

6.6 What is the procedure for estimating workhours from a factored estimate?

6.7 How are the workhours for setting equipment determined for a factored estimate?

6.8 What contributes to improved accuracy in making workhour projections for a preliminary estimate as compared to an order-of-magnitude or factored estimate?

6.9 What is required for the proper utilization of construction labor for a preliminary estimate?

6.10 What bulk accounts are frequently underestimated in preliminary estimates?

6.11 For a definitive estimate, what percentage of plant engineering should be completed?

6.12 For a definitive estimate, what percentage of major equipment items should have been quoted by vendors?

6.13 For a detailed estimate, what happens to labor productivity near the end of a project? Why?

6.14 a) What does labor planning require? b) What is the result of the detailed labor plan?

6.15 What is meant by craft and labor leveling?

6.16 Under what circumstance should overtime be used on a construction project?

7

Importance of the Capital Estimate

There is a growing tendency to place great emphasis on fancy computerized financial analysis. The truth is that all the computer printouts one can generate will not suffice if indeed the financial analysis is based on a poor capital estimate. That is, a financial analysis is a great tool when properly used, but can lead to a wrong conclusion if used improperly.

Generally, the overriding elements which make up operating costs are (1) raw material costs and (2) capital-related costs. This is not intended to indicate that other costs are unimportant, but rather to indicate that a properly executed capital estimate is important not only in having the correct capital number available, but also as a significant element of the estimated annual operating cost. Thus a poor estimate is double trouble because not only is the capital estimate incorrect, but also the operating cost, and these errors compound in the same direction. If the capital estimate is low, the operating cost estimate is low and vice versa. Therefore, it is imperative that the best estimate that can be made under the circumstances be made, regardless of the stage of project development. One should keep in mind that doing preliminary estimates of large, complex, integrated plants is a precarious occupation even for the experienced estimator. It is difficult to qualify the estimate to the extent needed without destroying credibility. However, it is the responsibility of the estimator(s) to inform management continually as concisely as possible as to the accuracy and stage of development of the capital estimate.

It is the responsibility of the estimator(s) to eliminate surprises for management; that is, management should *not* be shocked by being informed when the project is 70% completed that a project will overrun 25%, because they should have been forewarned of this possibility much earlier in the project development.

Therefore, keeping the capital estimate updated with continual improvement of accuracy is the primary concern of cost engineers on a construction project. This concern precipitates other numerous activities such as tracking bulk accounts, following progress on quotations, ascertaining equipment delivery dates, tracking field labor, getting a feel for labor productivity, etc. All such activities are necessary to control the final capital cost.

SCHEDULING

Scheduling is a most important activity on any construction project and in many cases prolonging the schedule for any reason can be catastrophic.

Sometimes, cost control engineers are referred to as cost and scheduling engineers. That does not mean that all cost control engineers are schedulers or that all schedulers are cost control engineers, but it does indicate that the cost control engineer cannot do his or her work without a project schedule.

The development of the schedule requires input from all of the key personnel on the project and is the basis for all construction activity. The schedule must be developed with consideration of the following items:

- Environmental considerations
- Sequence of events
    - Critical path method
    - Long lead time equipment
- Labor loading
- Peak construction force
- Labor and craft availability
- Leveled craft requirements

- Work breakdown structure
- Socioeconomic impacts
  - Schools
  - Transportation system, roads, etc.
- Input from all engineering disciplines
- Input from construction experts

Environmental considerations can be handled but do require planning and timely filing of permit applications. If one permit is omitted, the project can be delayed for years. For large capital-intensive projects, such delays after initiating field construction are absolutely intolerable. Therefore, it is to the owner's advantage to have an expert identify and initiate the procedures for obtaining the necessary permits at the time required. For example, if environmental monitoring is necessary, then a full year of baseline data must be collected in a prescribed manner to meet the guidelines of the cognizant regulatory agency or agencies. It is well to remember that all government work is done at government's convenience. A discussion of all aspects of environmental regulations would require several volumes, but here it suffices to say that improper handling will destroy a carefully worked out schedule.

Sequencing of events in a project schedule must be logical, and it is obvious that some activities must be completed before others can occur. For example, process engineering must be partially completed before design engineering can be initiated. The underground piping is normally completed before foundations are put in place. Some design engineering must be completed before procurement can be done, etc. The critical path method (CPM) and other methods have been proposed for analyzing a schedule. A properly worked out CPM model should result in a schedule that depicts the project sequencing of activities which gives the shortest schedule. Almost any schedule for a construction project should have some float built in. Float is extra time placed in the critical path to allow for unforeseen events such as bad weather (worse than normal), jurisdictional disputes among crafts on union jobs, and other construction activity disruptions.

Long lead-time equipment should be identified early in the project so that the design can be completed, the bid package prepared, and procurement initiated to assure timely arrival at the construction site. Failure to consider the delivery times required for all of the equipment and the bulk materials will result in chaos on the project.

The schedule must be labor loaded to facilitate the distribution of crafts and labor on the project and is the beginning of the work breakdown structure. That is, for all activities which appear on the schedule, the number of people required to perform the work must be estimated along with the time required to complete the task. Also, the number of laborers, craftspersons, etc., must be estimated to expedite craft leveling. One cannot reasonably expect to have 102 pipefitters one day, 2 the next day, and 52 the next. Craft leveling is an absolute requirement on a construction project.

The more complete the work breakdown structure, the better the labor estimates. Then the labor projections can be built up with reasonable assurance of acceptable accuracy for the entire project.

It is good practice to get as much input as possible from all of the project experts including managers, engineers, construction experts, etc., for scheduling. People do a much better job of meeting their own objectives; that is, people work much more diligently when they have had input to the project goals expressed by the schedule.

For large projects and for any project in a remote, sparsely populated area, the socioeconomic impacts on the local infrastructure must be considered. That is, the adequacy of the school system, roads, housing, and the effect on the local economy are all important aspects of scheduling. For example, some megaprojects in remote areas have required the construction of complete new towns and the importation of thousands of construction people. These are extreme examples, but in any situation, the availability of craftspersons and laborers must be addressed.

The schedule development parallels the project evolution starting with a simplistic, somewhat undefined schedule to the final detailed schedule showing the labor loading, work breakdown structure, and so on. It follows that the project schedule is the basis for the capital expenditure schedule and interest during construction and thus enhances the accuracy of the financial analysis. The schedule is, and should be, the center of attention on a construction project.

CONTINGENCY

Contingency deserves a separate discussion because of the rampant misunderstanding of what contingency on a project represents. As stated previously, it does not cover poor estimating or scope changes, but rather is an estimate of the accuracy considering the stage of development of the scope of work or the stage of development of the project. It follows that the contingency decreases as the project scope is more clearly defined. One should keep in mind that the contingency is not excess money in an estimate. It should be spent if indeed the project proceeds to fruition.

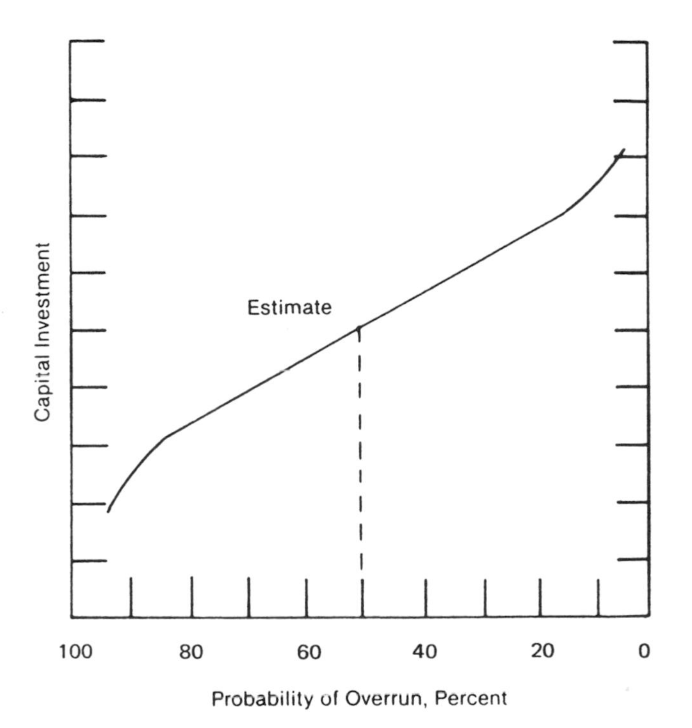

Figure 7.1 Typical probability curve. A preliminary estimate would include about 25% contingency; a lesser contingency would increase the probability of overrun; and a greater contingency would decrease the probability of overrun.

Contingency, regardless of calculating methodology, is a matter of judgment. There are several ways to calculate contingency, such as generating probability curves, etc., but in most engineering and construction contractor firms, management reserves the right to set the allowance.

A contingency should be related to some probability of overrun on the project, and the most common contingency gives a 50:50 probability of same. Obviously, if a 20% probability of overrun is desired, the contingency would be increased. Therefore, contingency may be calculated for several probabilities and presented in tabular form.

Probability curves take the general form shown in Figure 7.1. These curves break at about 16 and 84%. Thus the 85% probability of no overrun is normally the highest contingency quoted for a project.

QUESTIONS

7.1 What is one of the pitfalls of computerized financial analysis?
7.2 What are the overriding elements of operating costs?
7.3 What is the responsibility of estimator(s) to management?
7.4 What is the primary concern of cost engineers on a construction project?
7.5 What activities does this primary concern precipitate?
7.6 Why are environmental permits so important?
7.7 What should one always remember concerning government agency work?
7.8 What is a CPM?
7.9 What is float time in a schedule?
7.10 Why is craft leveling required on a construction project?
7.11 Why is it desirable to obtain scheduling input from all of the experts working on the project?
7.12 What is the center of attention on construction projects?
7.13 What is the rationale and purpose of a contingency?
7.14 Concerning contingency, what do most engineering and construction firms reserve the right to do?

8

Sources of International Cost Data*

INTRODUCTION

Industrial firms are increasingly multinational, and many companies now build and operate plants in several nations. Economic cooperation agreements between nations, such as the European Economic Community (EEC), the North American Free Trade Agreement (NAFTA), and the General Agreement on Tariffs and Trade (GATT) are encouraging further industrial globalization.

It is therefore becoming increasingly important for the cost engineer and cost estimator to be familiar with techniques for estimating costs in other countries and to be able to compare costs in different nations. The questions that are always being asked are: "What will this plant cost in the U.S.?"

* The information in this chapter is primarily based upon a paper of the same title by B. G. McMillan and K. K. Humphreys, *Transactions of the Association Française des Ingénieurs et Techniciens d'Estimation, de Planification et de Projets and Proceedings of the Eleventh International Cost Engineering Congress*, Paris, 1990.

"What about Australia, Germany, Norway, and Mexico?" and "Which location is the most attractive financially?"

A good cost engineer can readily develop the estimate for a plant in the cost engineer's home country, but such is not always the case for other nations. While cost engineers are very familiar with major sources of cost data in their own country, they are often unaware of useful sources of cost data and related information in other countries. This problem is exacerbated by lack of time to perform a proper search, publications in various languages, and lack of information about key factors that can impact the estimate for particular geographic locations.

The following discussion is far from complete but should provide some insight into the availability of international cost data and location factors. Included are:

1. Description of multicountry data sources for preliminary or conceptual cost studies
2. Notation of major database sources of unit price/cost data for specific countries
3. Presentation of some example data to illustrate the kind of resources available for international project studies

LIMITATIONS OF PUBLISHED DATA

The information sources described in this chapter are primarily periodical publications of various types. Published information must always be used with care. Every location factor or commonly available cost index has its own underlying method of construction, with its particular inherent components and weightings. It is vital for the estimator using such quick-estimate data to understand how it was created, and to recognize just what its limitations and applications are. Published data also is often inadequately explained and is frequently improperly dated. Date of publication is meaningless because the data may be months or years old and may require adjustment to current cost levels. Equipment cost data may or may not include ancillaries and/or transportation and installation costs, etc. Too often it seems that in the rush to complete the assignment, people will grasp any number they can find without fully understanding how it was derived, or what it represents.

With location factors, one must recognize that they generally reflect only the relative cost to replicate a facility exactly in another location. The factors do not consider cost effects which are introduced by site-unique conditions such as climate, earthquake and geological considerations, etc. If the design is not identical in both locations, the cost differences are not generally accounted for if location factors alone are used.

BACKGROUND SOURCES

It is beyond the scope of this book to discuss the many complex problems that the cost engineer must recognize in preparing estimates for international projects. Nevertheless, due to their importance, some mention of the available background literature is appropriate.

In order for a proper estimate to be made of a particular international project, it is vital that the cost engineer understand the conditions existing in the country or countries where a project is to be located. In a 1978 paper presented at the Fifth International Cost Engineering Congress, Utrecht, Netherlands, C. G. Walker (1) outlined the major economic system parameters to be evaluated as follows:

- Political
  - Stability
  - Attitude towards foreign investment
  - Type of bureaucracy
- Finance
  - Banking system
  - Insurance regulations
  - Tax system
  - Duties
- Legal system: the laws governing conduct of business and individual freedom
- Social system
  - Business ethics
  - Education
  - Language and religion
- Geography
  - Infrastructure and communication
  - Climate
- Industry
  - Capacity
  - Diversity
  - Efficiency

Table 8.1 Economic Basis of Cost: Levels at Which the Economic Systems Impact on Cost

Economic system	Area of influence				
	International	National	Local	Project	Company
Political	x	x	x		x
Financial	x	x		x	x
Legal		x		x	x
Social	x	x	x		x
Geographic		x	x	x	
Industrial	x	x	x		

Walker expressed the interactions of these factors in matrix form as shown in Table 8.1.

Many other authors (2–10) have discussed site- or country-specific factors that can impact the cost, schedule, and/or price for an international project. These include: local material quality/availability, labor availability, equipment availability, labor productivity, import duties, import licenses, local taxes, language, length of workweek, holidays, inflation, fluctuating exchange rates, religious customs, buy-local laws, shipping cycles, weather/climatic impacts, workforce level of education, logistics, workforce housing, and many other relevant factors. Regional variations of these factors within a country must also be expected, and remoteness/distance from major cities or supply centers can often aggravate the above problems even further.

A. Patrascu (11–13) has proposed pre-estimate survey checklists to help identify background concerns for foreign construction projects. These checklists delineate a large number of factors which must be considered, including those described by Walker (1).

For offshore projects, an excellent detailed checklist (14) is available from the Association of Cost Engineers (ACE), Lea House, 5 Middlewich Road, Sandbach, Cheshire CW11 9XL, United Kingdom. Based on North Sea oil field construction experience, this publication provides a very detailed breakdown for offshore work.

Many other practical references (e.g., 15–27) are useful in preparation for estimating international projects. However, background literature such as the publications mentioned here cannot replace pre-estimate site visits, proper contract development, and talking with others who have experience estimating work in the particular country or countries of interest. Review of appropriate literature, however, can help to ensure that all important factors for the project have been considered in developing the estimate.

To this point, this chapter has described the extreme complexity involved in costing out-of-country projects. The references mentioned, while far from being a complete list, should nevertheless make the cost engineer aware of most of the potentially important considerations which are unique to international work.

INTERNATIONAL LOCATION FACTORS

When little time is available or warranted to perform the type of background studies suggested above, and detailed design and engineering has not been completed, estimators must turn to published indexes, location factors, or other sources of relevant data for help. Table 8.2 shows a collection of sources for multicountry information.

The references in Table 8.2 are examples of multicountry data sources. For conceptual studies these may be useful, depending on what countries are of interest. Other references (e.g., 6, 35, and 36) may also be of help. Various reports by banks, governments, trade associations, etc., also exist if one is willing to search for them. An example is a report (37) that compares building costs in a number of countries to those in Malaysia. This special report was issued by a government task force and provides a handy reference for those comparing building costs in Pacific Rim countries.

Massa (3,4) has proposed development of international cost location factors based upon a weighting of 33.05% for a labor factor, 53.45% for an equipment and civil material factor, and 13.50% for an indirect and home office cost factor. He presents a detailed form for calculating these three factors and the composite factor for any given country referenced to U.S. Gulf Coast costs. Massa has also provided labor factors for many countries. His factors are presented in Table 8.3 along with a list of country location factors previously reported by Bridgewater (5,38). The Bridgewater factors are for complete chemical plants and are referenced to both the United Kingdom and the United States.

Table 8.2 International Data Sources: Multiple Country

Foreign Labor Trends (15)—periodic reports of labor trends and costs for specific countries—includes key labor indicators, information on unionization, labor availability, recent developments affecting the workforce, etc. Each report covers one specific country and is prepared by the American Embassy staff in that particular country.

Hanscomb/Means International Construction Cost Intelligence Report (16) —newsletter—provides comparative building construction cost information for many countries. Feature articles discuss costs and construction conditions in specific countries and geographical regions.

Engineering Costs and Production Economics (28)—journal—indexes of erected costs of plants for: Belgium, Denmark, France, Germany, Italy, Netherlands, UK, Australia, Canada, Japan, Norway, Sweden, US; separate indexes for labor and steel sections (heavy), steel rebar, steel plate, and portland cement; index weightings: 7% civil and building materials, 63% mechanical and electrical equipment and materials, including engineering and procurement, and 30% field construction costs.

U.N. Monthly Bulletin of Statistics (29)—includes a variety of production, trade, financial, commodities, construction, wage, and other cost/price indexes and statistics for about 190 countries.

Engineering News-Record (30)—weekly magazine—primarily U.S. and Canada, variety of commercial/industrial construction and builders' indexes plus materials prices and labor rates; substantial North American data featured in "Quarterly Cost Roundup" issues; "World Parameter Costs" column features building costs for one or two countries per quarterly issue.

Costos de Construccion Pesada y Edificacion (Heavy Construction and Building Costs) (31)—cost estimating database on building, industrial and heavy construction reflecting costing in Mexico and other Latin American countries, e.g., Bolivia, Venezuela, Panama, Brazil, and Chile.

Spon's Architects' and Builders' Price Book (32)—current edition—in addition to 3 unit cost books published by Spon's for the UK, the *A&B Price Book* contains a European section for tendering and costs of labor and materials in 13 European countries. *Spon's European Construction Costs Handbook* provides coverage of 25 countries in

Europe plus the United States and Japan. Country listings include country descriptions, contract procedures, inflation rates, currency exchange rates, approximate estimating rates per m^2 of different buildings, unit materials and labor costs, and comparative European construction cost data from Eurostat. *Spon's Asia Pacific Construction Costs Handbook* provides similar coverage of 13 countries in Asia. The two handbooks are also distributed by the R.S. Means Co. (33) in North America.

R.S. Means Co. (33)—various cost books published annually for building and industrial construction in the U.S. and Canada.

Richardson Engineering Services (34)—unit cost database published for U.S. and Canada building and general construction; special estimating set for process plant construction; international estimating manual.

Table 8.3 Sample Location Factors*

	Massa's (3,4) Labor Productivity Factors	Bridgewater's (38) Factors for Chemical Plants	
Location	US = 1.0	UK = 1.0	US = 1.0
Algeria	1.82	—	—
Argentina	2.00 (1.30–2.60)	—	—
Australia	1.20 (0.96–1.45)	1.4	1.3
Austria	1.60 (1.57–2.10)	1.1	1.0
Belgium	1.14 (1.14–1.50)	1.1	1.0
Canada, East	1.14 (1.08–1.17)	—	—
Canada, West	1.07 (1.02–1.11)	—	—

Table 8.3 (*Continued*)

	Massa's (3,4) Labor Productivity Factors	Bridgewater's (38) Factors for Chemical Plants	
Location	US = 1.0	UK = 1.0	US = 1.0
Canada	—	1.25	1.15
Central Africa	—	~2.0	~2.0
Central America	—	1.1	1.0
Chile	2.70 (2.00–2.90)	—	—
China			
(imported element)	—	1.2	1.1
(indigenous element)	—	0.6	0.55
Colombia	3.05	—	—
Denmark	1.28 (1.25–1.30)	1.1	1.0
Egypt	2.05	—	—
Finland	1.28 (1.24–1.28)	1.3	1.2
France	1.52 (0.80–1.54)	1.05	0.95
Germany (West, prior to reunification)	1.20 (1.00–1.33)	1.1	1.0
Ghana	3.50	—	—
Greece	1.49	1.0	0.9
India	4.00 (2.50–10.0)	—	—
(imported element)	—	2.0	1.8
(indigenous element)	—	0.7	0.65
Iran	4.00	—	—

Table 8.3 (*Continued*)

Location	Massa's (3,4) Labor Productivity Factors	Bridgewater's (38) Factors for Chemical Plants	
	US = 1.0	UK = 1.0	US = 1.0
Iraq	3.50	—	—
Ireland	—	0.9	0.8
Italy	1.48 (1.10–1.48)	1.0	0.9
Japan	1.54 (1.00–2.00)	1.0	0.9
Malaysia	—	0.9	0.8
Mexico	1.56 (1.54–3.15)	—	—
Middle East	—	1.2	1.1
Netherlands	1.25 (1.25–1.60)	1.1	1.0
Newfoundland	—	1.3	1.2
New Zealand	—	1.4	1.3
Nicaragua	2.67	—	—
Nigeria	2.22	—	—
North Africa			
(imported element)	—	1.2	1.1
(indigenous element)	—	0.8	0.75
Norway	1.23	1.2	1.1
Philippines	2.86	—	—
Portugal	1.66	0.8	0.75
Puerto Rico	1.54	—	—
Singapore	4.00	—	—

Table 8.3 (*Continued*)

Location	Massa's (3,4) Labor Productivity Factors	Bridgewater's (38) Factors for Chemical Plants	
	US = 1.0	UK = 1.0	US = 1.0
South Africa	1.58	1.25	1.15
South America (N)	—	1.5	1.35
South America (S)	—	2.5	2.25
Spain	1.74	—	—
(imported element)	—	1.3	1.2
(indigenous element)	—	0.8	0.75
Sri Lanka (Ceylon)	3.50	—	—
Sweden	1.18 (1.10–1.20)	1.2	1.1
Switzerland	—	1.2	1.1
Taiwan	1.52 (1.52–7.20)	—	—
Thailand	2.82	—	—
Turkey	2.32	1.1	1.0
United Kingdom	1.53 (0.70–2.46)	1.0	0.9
United States	1.00	1.1	1.0
Venezuela	2.00	—	—

* Note: Increase chemical plant factor by 10% for each 1000 miles or part of 1000 miles that the new plant location is distant from a major manufacturing or import center or both. When materials or labor, or both, are obtained from more than a single source, pro-rate the appropriate factors. Factors do not consider investment incentives.

COUNTRY COST INDEXES

Cost indexes are valuable proven tools for adjusting costs for changes over time and, in combination with appropriate location factors, can facilitate development of conceptual estimates. The American Association of Cost Engineers' *Cost Engineers' Notebook* (39) describes 28 indexes and index sources for U.S. and Canadian costs and 27 indexes and index sources for many other countries, including Australia, Brazil, England, France, Italy, Japan, Mexico, the Netherlands, New Zealand, and Germany. A description of each index or index source is given in Table 8.4.

In addition, *Engineering Costs and Production Economics* (28) regularly publishes indexes of erected costs of plants for Belgium, Denmark, France, Germany, Italy, the Netherlands, United Kingdom, Australia, Canada, Japan, Norway, Sweden, and the United States. A description of these indexes is given in Table 8.2.

COST DATA SOURCES

Various compilers and publishers maintain databases of costs which form the basis for their various cost publications. A few of these are listed in Table 8.5. Beyond these, an invaluable source of information and costs are the various cost engineering and project management societies throughout the world. Their members and employing firms are the most valuable network of international cost information that exists anywhere.

At the time of this writing, the 19 member societies of the International Cost Engineering Council (ICEC) represented, either directly or through their branches and chapters, over 40 countries.

The International Project Management Association (INTERNET) similarly had about 19 member societies representing many nations. A list of the members of ICEC and INTERNET is given in Table 8.6. Because many of these societies do not have permanent headquarters offices, it is suggested that ICEC or INTERNET be contacted for current addresses of each society if they are not otherwise available.

SUMMARY

The best summary for this chapter is from John E. Barry's paper, "Ten Commandments of International Cost Engineering," presented at the 1993 Annual Meeting of AACE International (25):

Table 8.4 Country Indexes*

Country/Index	Comments
Australia	
Building Cost Index	Construction labor and material costs (40)
Construction Cost Index	Weekly earnings; building and nonbuilding materials (41)
Australia Builder	Price information on raw materials (42)
Cordells Building Cost Book	Price information for non-residential building (43)
Monthly Summary of Statistics	Manufacturing articles: materials, building/non-building materials, metallic materials, wage rates; indexes (44)
Brazil	
Revista de Precos	Cost/price indexes for residential and non-residential construction; Portuguese (45)
Boletim de Custos	Cost/price indexes for residential and non-residential construction; Portuguese (46)
A Construcao	Residential and non-residential construction and project costs; Portuguese (47)
Conjuntura	Indexes for industrial machinery and equipment; residential and non-residential construction material; Portuguese (48)
NTC—Associacao Nacional dos Transportadores de Cargas	Transportation rates for industrial materials; Portuguese (49)
Canada	
Statistics Canada	Variety of cost/price indexes for construction costs and capital expenditures; English and French (50)
England	
A.C.E. Indices of Erected Plant Costs	Indexes of erected cost of typical process plants (51)

Table 8.4 (*Continued*)

Country/Index	Comments
England (*Continued*)	
Price Index Numbers for Current Cost Accounting	Indexes for non-residential construction, machinery, equipment (52)
France	
Index Coefficients	General index for building and machinery replacement costs (53)
Germany	
Fachserie 17: Preise	Indexes for general machinery and building construction; German (64)
Italy	
Indicatori Mensili	Government statistics and indexes; Italian (54)
Index	Costs and indexes for various industries; Italian (55)
Prezzi Informativi Delle Opere Edili in Milano	Indexes for residential and non-residential construction, consumer prices, raw materials, and wage rates; Italian (56)
Japan	
Construction Price Indexes by Year, Price Indexes of General Machinery and Equipment	Non-Life Insurance Institute publication for use by industrial insurers in Japan; English (57)
MRC Monthly Standard Building Cost Indexes and Unit Price Data Bulletin	Cost indexes of many components of construction and certain building types; Japanese (58)
Mexico	
Cifras de la Construccion	General construction index; Spanish (59)
Indice Nacional de Precios al Consumidor	Consumer price index; Spanish (60)

Table 8.4 (*Continued*)

Country/Index	Comments
The Netherlands	
Die Workgroep Begrotings Problemen in de Chemische Industrie (WEBCI)	Unit prices for chemical plant construction in the Netherlands; Dutch (61)
New Zealand	
Ministry of Works and Development Construction Cost Index	Weighted construction cost index (62)
Monthly Abstract of Statistics	Variety of labor force indexes by indusrty; residential construction, wage rates, other (63)
United States	
General purpose indexes:	
**ENR* Construction	20–city construction cost
**ENR* Building	20–city building cost
*U.S. Dept. of Commerce	composite (65)
*BuRec	general building
*Construction Industry Institute	construction price (66)
*Factory Mutual	industrial building
*Handy Whitman	building construction (67)
*Lee Saylor, Inc.	material/labor
*R.S. Means	construction cost (33)
Selling price indexes, building:	
*Fru-Con Corp.	industrial
*Lee Saylor, Inc.	subcontractor
*Turner	general building (68)
*Smith, Hinchman & Grylls	general (69)

Table 8.4 (*Continued*)

Country/Index	Comments
Valuation indexes:	
*Boeckh	20–city commercial/manufacturing (70)
*Marshall & Swift	industrial equipment (71–72)
Special Purpose:	
*Nelson-Farrar	refinery cost inflation index
**Chemical Engineering*	plant cost (72)
*Federal Highway Construction Bid Price	(73)
*Handy Whitman	public utility construction
*BuLabor Statistics Consumer Price Index	(74)
*BuLabor Statistics Producers Price Index	(65)
AED Average Rental Rates for Construction Equipment	(75)

* Commonly published in *Engineering News-Record* (30) "Quarterly Cost Roundup" issues.

Table 8.5 Country Databases for Project Costs

Country	Source
Australia	*Building Economist* quarterly (76)
	Cordells Building Cost Book (43)
Brazil	*A Construcao* (47)
	Conjuntura, monthly (48)
Canada	R.S. Means, various books (33)
	Richardson Engineering books (34)
Mexico	*Costos de Construccion Pesada y Edificacion* (31)
Netherlands	*WEBCI Price Book* (61)
United Kingdom	E & FN Spon, various books (32)
United States	R.S. Means, various books (33)
	Richardson Engineering books (34)

Table 8.6 1995 Member Societies of the International Cost Engineering Council (ICEC) and the International Project Management Association (INTERNET)

Society	Country	ICEC member	INTERNET member
AACE-CANADA, The Canadian Association of Cost Engineers	Canada	Yes	No
AACE International (AACE) (chapters in Canada, Saudi Arabia, Japan, Australia, South Africa, Puerto Rico, and Norway)	United States	Yes	Yes*
Association of Cost Engineers (ACE) (chapter in Hong Kong)	United Kingdom	Yes	No
Association Française des Ingénieurs et Techniciens d'Estimation, de Planification et de Projets (AFITEP) (chapters in Belgium and Switzerland)	France	Yes	Yes
Association of Project Managers	United Kingdom	No	Yes
Association of South African Quantity Surveyors (ASAQS)	South Africa	Yes	No
Associazione Nazionale di Implantistica Industriale (ANIMP)	Italy	No	Yes
Associazione Italiana di Ingegneria Economica (AICE)	Italy	Yes	No
Australian Institute of Quantity Surveyors (AIQS)	Australia	Yes	No

Table 8.6 (*Continued*)

Society	Country	ICEC member	INTERNET member
Cost Engineering Association of Southern Africa (CEASA)	11 countries in southern Africa	Yes	No
ČSNK INTERNET	Czech Republic	No	Yes
Egyptian Society of Engineers, Management Engineering Society (MES)	Egypt	No	Yes
Federacion Panamericana de Ingeniería Económica y de Costos (FEPIEC)	many Pan-American countries	Yes	No
Gépipari Tudományos Egyesület / Mûszaki Költségtervezõ Klub (GTE/MKK)	Hungary	Yes	No
Gesellschaft für Projektmanagement (GPM)	Germany	No	Yes
Hellenic Project Management Institute	Greece	No	Yes
Instituto Brasileiro de Engenharia de Custos (IBEC)	Brazil	Yes	No
Irish Project Management Institute	Ireland	No	Yes
Korean Institute of Project Management and Technology (PROMAT)	South Korea	Yes	No
Magyar Fovallalkozok Szovetseye	Hungary	No	Yes
Nederlandse Stichting Voor Kostentechniek (DACE)	Netherlands	Yes	No
Netvaerket for Dansk Projektledelse (NDPM)	Denmark	Yes	Yes

Table 8.6 (*Continued*)

Society	Country	ICEC member	INTERNET member
Norsk Forening for Prosjektledelse (NFP)	Norway	Yes	Yes
Project Management Institut Nederland	Netherlands	No	Yes
Projektitoimintayhdistys (PTY)	Finland	Yes	Yes
Projektmanagement Institut INTERNET Austria	Austria	No	Yes
Russian Project Management Institute	Russia	No	Yes
Slovenian Project Management Institute	Slovenia	No	Yes
Sociedad Mexicana de Ingeniería Económica, Financiera y de Costos (SMIEFC)	Mexico	Yes	No
Svenskt ProjektForum (SPMS)	Sweden	Yes	Yes
Swiss Society for Project Management (SPM)	Switzerland	No	Yes
Ukranian Project Management Institute	Ukraine	No	Yes
Verkefnastjórnunarfélag Íslands (PMAI)	Iceland	Yes	Yes

ICEC Secretariat address:
23 Hidden Lake Drive
Granite Falls, NC 28630
USA
Phone: 1–704–728–5287
Fax: 1–704–728–0048

INTERNET Secretariat address:
Claridenstrasse 36
CH-8002 Zurich
SWITZERLAND
Phone: 41–1–28–78–181
Fax: 41–1–28–78–100

* Corporate member

> The reliability of available data is usually suspect, so test, test, and retest it. Communication problems are difficult across cultures, and sometimes language differences can cause misunderstandings when soliciting data. The use of international factors is a session all its own. A few basic principles for guidance include keeping in mind that a time or place factor means nothing without an exchange rate and date and that a base city must be identifiable, since variation can exist within your home country.

The paper concluded with Barry's "Ten Commandments for Worldwide Cost Engineering," which have appeared in several of his articles over the years. They are a fitting conclusion to this chapter as well; they are:

1. Thou shalt not begin an international cost engineering assignment without preparing for the differences in culture and protocol.
2. Thou shalt not ignore investment objectives of the target country's government.
3. Thou shalt not look at building your own facilities as the only way to enter business in a country.
4. Thou shall not use biased estimated scope that does not reflect technical, cultural, legal, and climate differences.
5. Thou shalt not ignore what equipment must be imported and the impact on cost and schedule.
6. Thou shalt not accept as gospel cost data for other countries without thoroughly checking for understanding and testing for reasonableness.
7. Thou shalt not ignore productivity, weather, religious practices, and construction methods when calculating labor cost.
8. Thou shalt not ignore the additional risks associated with cost and schedule on international projects.
9. Thou shalt not forget that AACE and ICEC members are valuable resources who are capable and willing to help.
10. Thou shalt not ignore the previous nine commandments.

PROBLEMS AND QUESTIONS

8.1 Why is it important for the cost engineer or estimator to be familiar with techniques for estimating costs in other countries?

8.2 What do location factors reflect?

8.3 What site- or country-specific factors impact the cost, schedule, and/or price for an international project?
8.4 If a chemical plant construction project is estimated to cost \$10,000,000 in the United States, what would be the approximate cost of the same project in Central Africa? in New Zealand?
8.5 What is ICEC? What is INTERNET?
8.6 If a project in Greece requires 25,000 workhours of labor, what would the approximate labor requirement be for the same project in the United States? in Colombia? in Venezuela?

REFERENCES AND INFORMATION SOURCES

1. Walker, C.G., "Estimating Construction Costs Abroad," *Transactions of the Fifth International Cost Engineering Congress*, Paper A.10, Dutch Association of Cost Engineers, The Hague, Netherlands, 1978.
2. Cran, J., "A Cost and Location Index Grid," *AACE Transactions*, Paper F.2, American Association of Cost Engineers, Morgantown, WV, 1984.
3. Massa, R.V., "A Survey of International Cost Indexes, A North American Perspective," 1st European Cost Engineering Forum, International Cost Engineering Council, Oslo, Norway, 1985.
4. Massa, R.V., "International Composite Cost Location Factors," *AACE Transactions*, Paper F.1, American Association of Cost Engineers, Morgantown, WV, 1984.
5. Humphreys, K.K., *Jelen's Cost and Optimization Engineering*, 3rd ed., McGraw-Hill Book Co., New York, 1991, pp. 423–428.
6. Kharbanda, O.P., *Process Plant and Equipment Cost Estimation*, Craftsman Book Co., Carlsbad, CA, 1979.
7. Richter, I.E., *International Construction Claims: Avoiding and Resolving Disputes*, McGraw-Hill Book Co., New York, 1983.
8. Stallworthy, E.A., *International Construction and the Role of Project Management*, Gower Press, Brookfield, VT, pp. 128–133, 1985.
9. Kharbanda, O.P. and E.A. Stallworthy, *How to Learn from Project Disasters*, Gower Publishing Co., Hampshire, UK, 1983.
10. Kerzner, H., *Project Management: A Systems Approach to Planning, Scheduling and Control*, Van Nostrand Reinhold, New York, 1989, pp. 895–920.
11. Patrascu, A., *Construction Cost Engineering*, Craftsman Book Co., Carlsbad, CA, 1978, p. 55.
12. Patrascu, A., *Construction Cost Engineering Handbook*, Marcel Dekker, Inc., New York, 1988.
13. Patrascu, A., "Pre-estimating Survey. The International Aspect," *AACE Transactions*, Paper B.A, 1979, and *Cost Estimating: Concepts and Approaches*,

AACE Technical Monograph Series, CE-1, 1989, American Association of Cost Engineers, Morgantown, WV, 1989.

14. *Estimating Check List for Offshore Projects*, Association of Cost Engineers, Sandbach, Cheshire, United Kingdom, 1982.
15. *Foreign Labor Trends*, U.S. Department of Labor, Bureau of International Affairs, Washington, DC. Periodic reports of labor trends in various countries.
16. *Hanscomb/Means International Construction Cost Intelligence Report* (newsletter), Hanscomb Associates, 1175 Peachtree St., N.E., Atlanta, GA 02356.
17. Smith, K.A., "Forecasting Construction Costs for International Projects," *AACE Transactions*, Paper C.1, American Association of Cost Engineers, Morgantown, WV, 1978.
18. Grinberg, M., "Planning and Scheduling an International Project," *AACE Transactions*, Paper H.7, American Association of Cost Engineers, Morgantown, WV, 1981.
19. Milot, M., "Third World Estimating in Project Management," *AACE Transactions*, Paper O.5, American Association of Cost Engineers, Morgantown, WV, 1984.
20. Sinclair, D.N., *Estimating for Abnormal Conditions*, Industrial Press, Inc., New York, 1989.
21. Ruddock, L., "Productivity in the UK Construction Industry," *AACE Transactions*, Paper L.1, American Association of Cost Engineers, Morgantown, WV, 1991.
22. Hinojosa, L.C. and R. Vazquez del Mercado, "Mexico Infrastructure Projects Cost Engineering Support," *AACE Transactions*, Paper G.3, American Association of Cost Engineers, Morgantown, WV, 1992.
23. Watson, R.B., "Cost Control of Engineering Sources in Buildings," (UK) *AACE Transactions*, Paper E.5, American Association of Cost Engineers, Morgantown, WV, 1992.
24. Hinojosa, L.C. and R. Vazquez del Mercado, "Parametric Cost Estimating of Building Construction—US, Canada, and Mexico," *AACE Transactions*, Paper K.2, American Association of Cost Engineers, Morgantown, WV, 1993.
25. Barry, J.E., "Ten Commandments of International Cost Engineering," *AACE Transactions*, Paper L.4, AACE International, Morgantown, WV, 1993.
26. "Symposium INT" (Eight papers on international cost estimating and cost management), *AACE Transactions*, AACE International, Morgantown, WV, 1994.
27. *Transactions of the 13th International Cost Engineering Congress*, Association of Cost Engineers, Sandbach, Cheshire, United Kingdom, 1994.
28. *Engineering Costs and Production Economics* (quarterly journal), Elsevier Scientific Publishing Co., Amsterdam.
29. *United Nations Monthly Bulletin of Statistics*, United Nations, New York.
30. *Engineering News-Record* (weekly magazine), McGraw-Hill Book Co., New York.
31. Varela, Leopoldo, *Costos de Construccion Pesada y Edificacion*, Compubras, S.A., Anaya Monroy 168, Mexico DF 03590, MEXICO.
32. E & FN Spon, Ltd., 2–6 Boundary Row, London SE1 8HN, United Kingdom.

33. R.S. Means Co., 100 Construction Plaza, PO Box 800, Kingston, MA 02364–0800.
34. Richardson Engineering Services, 1742 S. Fraser Dr., PO Box 9103, Mesa, AZ 85214–1903.
35. Woodward, P.N., *Oil and Labor in the Middle East*, Praeger Publishing Co., One Madison Ave., New York, 10010.
36. Bu-Bshait, K.A. and Manzanera, I., "A Look at Planning & Scheduling in the Middle East," *AACE Transactions*, Paper D.6, American Association of Cost Engineers, Morgantown, WV, 1989.
37. Fong, C.K., ed., "Report on the Cost Competitiveness of the Construction Industry in Singapore," Construction Industry Development Board, 133 Cecil Street #09–01/02, Keck Seng Tower, Singapore 0106.
38. Bridgewater, A.V., "International Construction Cost Location Factors," *Chemical Engineering, 86*, p. 5 (Nov. 5, 1979).
39. *Cost Engineers' Notebook*, American Association of Cost Engineers, Morgantown, WV, 1989.
40. Building Cost Index, Australian Institute of Quantity Surveyors, National Surveyors House, 27–29 Napier Close, Deakin, ACT 2600, Australia.
41. *Construction Cost Index*, Amax Australia Ltd., 200 St. George's Terrace, Perth, WA 6000, Australia.
42. *Australian Builder*, Australian Builder Publishing Co. Pty. Ltd., 332 Albert Street, East Melbourne, Victoria 3002, Australia.
43. *Cordells Building Cost Book*, Cordells Building Publication, PO Box 96, East Melbourne, Victoria 3002, Australia.
44. *Monthly Summary of Statistics*, Australian Bureau of Statistics, PO Box 10, Belconnen, ACT 2616, Australia.
45. *Revista de Precos*, Av. N.S. de Capacabana, 749 Gr 801, Rio de Janeiro, Brazil.
46. *Boletim de Custos*, Rua Dana Mariana No. 2, Botafoga, Rio de Janeiro, Brazil.
47. *A Construcao*, Rua Anhaia 964, CEP 01130, Sao Paulo, Brazil.
48. *Conjuntura*, Instituto Brasileiro de Economia, Fundacao Getulio Vargas, Praia de Botafogo 188, Rio de Janeiro, Brazil.
49. NTC—Associacao Nacional dos Transportadores de Cargas, Ave Beira Mar 262, Gr 301–302, Rio de Janeiro, Brazil.
50. Statistics Canada, Prices Division, Ottawa, Ontario, Canada K1A 0T6.
51. A.C.E. Indices of Erected Plant Costs, Association of Cost Engineers, Ltd., Lea House, 5 Middlewich Road, Sandbach, Cheshire, CW11 9XL, England.
52. *Price Index Numbers for Current Cost Accounting*, Department of Industry, Economics and Statistics, Div. 4A, Sanctuary Buildings, 16–20 Great Smith Street, London SW1P 3D8, England.
53. *Index Coefficients*, Roux S.A., 51 Ampere, 75017, Paris, France.
54. *Indicatori Mensili*, Instituto Centrale di Statistica, Via Cesare Balbo 16–00100, Roma, Italy.
55. *Index*, Centro Statistica Aziendale S.r.l., Via A. Baldesi 20, 50131, Firenze, Italy.

56. *Prezzi Informativi Delle Opere Edili in Milano*, Camera di Commercio Industria Artigianato E, Agricoltura di Milano, Directtore Redazione Amministrazione, Via Meravigli 9/B1S, 20123 Milano, Italy.
57. *Construction Price Indices by Year*, Price Indices of General Machinery and Equipment, Non-Life Insurance Institute of Japan, 6–5, 3–Chome, Kanda, Surugadai, Chiyoda-ku, Tokyo, Japan.
58. *MRC Monthly Standard Building Cost Indexes and Unit Price Data Bulletin*, Management Research Society for Construction Industry, Japan.
59. *Cifras de la Construccion*, Fuente Camara Nacional de la Industria de la Construccion, c/o Sociedad Mexicana de Ingenieria Economica, Financiera y de Costos, Anaxagoras 17–101, Col. Navarte, 03020 Mexico DF, Mexico.
60. *Indice Nacional de Precios Al Consumidor*, Fuente Banco Nacional de Mexico, c/o Sociedad Mexicana de Ingenieria Economica, Financiera y de Costos, Anaxagoras 17–101, Col. Navarte, 03020 Mexico DF, Mexico.
61. *Dutch Association of Cost Engineers Prijzenboekje*, Nederlandse Stichting voor Kostentechniek/Stichting Nederlandse Apparaten voor de Processindustrie, PO Box 443, 2260 AK Leidschendam, Netherlands.
62. Ministry of Works and Development Construction Cost Index, Ministry of Works & Development, PO Box 12 041, Wellington North, New Zealand.
63. *Monthly Abstract of Statistics*, Department of Statistics, Aorangi House, 85 Molesworth Street, Wellington, New Zealand.
64. *Fachserie 17: Preise*, Verlag W. Kohlhammer GmbH, Abt. Veroffentlichungen des Statistischen Bunde Samtes, Philipp-Reis Strasse 3, Postfach 42 11 20, D-6500 Mainz 42, Hechtsheim, Germany.
65. Bureau of the Census, Construction Statistics Division, U.S. Dept. of Commerce, Washington, DC 20233.
66. Construction Industry Institute, 3208 Red River St., Suite 300, Austin, TX 78705–2650.
67. Whitman, Requardt & Associates, 2315 St. Paul Street, Baltimore, MD 21218.
68. Turner Construction Co., 150 E. 42nd Street, New York, 10017.
69. Smith, Hinchman & Grylls, Inc., 455 W. Fort Street, Detroit, MI 48226.
70. Boeckh Division, American Appraisal Associates, Inc., 525 E. Michigan St., Milwaukee, WI 53201.
71. Marshall & Swift, Inc., 1617 Beverly Blvd., Los Angeles, CA 90026.
72. *Chemical Engineering* (bi-weekly magazine), McGraw-Hill, Inc., 1221 Avenue of the Americas, New York, NY 10020.
73. U.S. Dept. of Transportation, Federal Highway Administration, Interstate Management Branch, HNG-13, Washington, DC 20590.
74. U.S. Dept. of Labor, Bureau of Labor Statistics (BLS), Washington, DC 20212.
75. Associated Equipment Distributors (AED), 615 West 22nd Street, Oak Brook, IL 60523.
76. *The Building Economist*, Australian Institute of Quantity Surveyors, National Surveyors House, 27–29 Napier Close, Deakin, ACT 2600, Australia.

Part II

Operating and Manufacturing Cost Estimates

9

Operating and Manufacturing Cost Estimating
An Overview

The operating or manufacturing cost determination requires a staffing plan or at least an estimate of the number of operators required per shift. The method used depends upon the relative progress of the overall project; that is, there is little to be gained by working out an elaborate staffing plan for a preliminary capital estimate.

Items to be considered in the estimate are listed in Table 9.1. The individual items shown in this table will be briefly discussed in this chapter in the same sequence as listed on the table. A more detailed discussion is provided in Chapter 10.

As stated earlier, much of the information necessary to calculate the estimated annual operating cost is generated from the work done to support the capital estimate. This is particularly true for a preliminary estimate.

The data necessary to calculate the raw materials and utility costs come from the utility balance, the materials balance, the thermal balance, and the buildup of catalyst and chemicals cost, all suggested as a part of the capital cost determination (see Chapter 2). The fuel requirement would be deter-

Table 9.1 Components of Annual Operating Cost

Direct costs:
- Raw materials and utilities
  - Fuel: Units per hour × hours per year × $ per unit
  - Power: kW/hr × hours per year × $ per kW
  - Steam: M lb per hour × hours per year × $ per M lb
  - Feed (raw material)
  - Annual catalyst and chemicals (from list)
- Direct labor
  - Direct labor: WH per day × $ per WH × days per year
  - Direct labor: Supervision × % of direct labor
- Plant maintenance
  - Maintenance labor
  - Maintenance labor supervision
  - Maintenance material
  - Contract maintenance

Payroll overhead (% of payroll)

Operating supplies

Indirect costs:
- Administration and general overhead

Fixed costs:
- Local taxes and insurance
- Depreciation

mined using both the materials and thermal balances, power requirement from the utility balance, steam from the utility balance (if purchased and not generated on site), raw materials requirements from the materials balance, and catalysts and chemicals from the catalyst and chemicals list.

The direct labor for a preliminary estimate is often based on the estimated number of operators per shift. This estimate can be made from historical patterns of similar plants or installations, if available. If not, an estimate can be made from a careful study of the flowsheets. Probably the best estimates come from experienced operations managers and engineers. In estimating, getting the best information from the best available source rarely costs more than poor estimating.

For a preliminary estimate the supervision is normally taken as about 15–20% of the direct labor. After the plant staffing plan is completed, the supervision is based on the estimated number of supervisors and on the annual salary projections.

The plant maintenance for a preliminary estimate is usually taken as a percentage of the depreciable capital cost. The literature gives ranges from 3–12% of the capital number; however, care must be exercised in choosing the factor. Small complicated plants would tend toward the high end of the range and large simple plants would tend toward the low end of the range, but this generalization is not always valid. For example, petroleum refineries in the United States average about 5% and are certainly complex processing systems. It follows that experienced personnel and historical data are the best sources of information on which to base the estimated factor.

Payroll overhead can only be determined from each individual operating company; however, the range for payroll overhead is 25–40% of the payroll, with an average near 30%. Payroll overhead can be determined exactly and thus improves in accuracy as the definition of the payroll improves.

Operating supplies include a multitude of disposable materials from hand soap to writing pads and is normally determined in a preliminary estimate as a percentage of the payroll which can range from as low as a few percent to 20% or more.

The indirect costs in a preliminary estimate may be estimated in several ways as described in Chapter 10. The indirect costs include all supervision above first-line supervisors, plant administration, laboratory support, safety and health, quality assurance, plant security, and any other support group which might be required for a specific installation. As the staffing plan is developed, the factor used is improved and then replaced with numbers generated from the staffing plan tabulation. The factor can vary widely depending upon how much of what is supplied for support. Also, an allowance for company support should be included if the installation is owned by a large corporation.

The fixed costs are capital-related costs and, thus, the capital estimate must be completed before the fixed costs can be calculated. The local taxes and insurance costs are site-specific costs and are determined from local tax rates and insurance premiums. The United States national average for such costs is about two percent. The tax laws dictate how depreciation is taken.

Table 9.2 Typical Electrical Summary

Item no.	Service	Operating kW	2-train total kW
Pumps			
17G-101 A/B	Fractioner feed	192	384
17G-102 A/B	Heavy distillate	151	302
17G-103 A/B	Flush oil pumparound	77	154
17G-104 A/B	Mid-dist. pumparound	129	258
17G-105 A/B	Heavy naphtha	86	172
17G-106 A/B	Light naphtha	30	60
17G-107 A/B	Sour water	14	28
17G-108 A/B	Mid-dist. product	28	56
17G-109 A/B	Flush oil product	25	50
17G-110 A/B	Desalter feed	75	150
17G-111 A/B	Aftercooler separator	1	2
	Subtotal	808	1616
Compressors			
17K-101 A/B	Vent gas compressor	366	732
	Subtotal	366	732
Fans			
17F-101	Combustion air	58	116
17F-101	Flue gas	60	120
	Subtotal	118	236

Theoretically, one has options as to which method of calculation is used. These options are discussed in detail in Chapter 10. However, one should consult with a tax expert as to viable options since tax laws are not static and do change from time to time.

Also, note that there is no mention of return on investment in Table 9.1. In almost all cases, a discounted cash flow rate of return is calculated on a project and both depreciation (non-cash item) and return on investment are accounted for in the financial analysis.

Forms used to present estimated annual operating costs reflect only minor changes as the project progresses, but the information upon which the operating cost calculations are based does change gradually to a more finite base. For example, the electrical power requirements in the later stages of the plant would be based on a complete motor list (e.g., see Table 9.2), total plant reliability, consideration of in-plant distribution losses, and reduction of the average power load to reflect contingencies normally put on data sheets. In a preliminary estimate the power would be calculated on installed horsepower corrected using an average power factor. It follows that all of the utility and raw material requirements would be better defined in the latter stages of a project.

The greatest improvement is that in the latter stages of a project a complete staffing plan is completed for the operation of the plant; therefore, all of the labor-related and management-related items are substantiated. The staffing plan should include an average contract maintenance work force. Also, the payroll overheads should be well defined, which leaves only maintenance materials and operating supplies to be factored. Since the labor is defined, the factors for operating supplies and maintenance materials are improved.

Table 9.3 is an example of part of a staffing plan for plant operations. Please note that the salaried and shift workers are treated differently. That is, the table reflects the fact that 1.4 shift workers are required to cover one shift for a full 7 days per week operation and that 5-day per week jobs are charged 260 days per year.

The staffing plan includes all plant personnel. The example given is taken from such a plan for a megaplant project; however, the same general organization may be used for any size plant.

When the project is near completion and the capital investment requirement has been defined, then the fixed costs can be accurately determined. Late in the project local taxes and insurance rates are known, and many estimators use straightline depreciation to complete the operating cost esti-

Table 9.3 Typical Plant Staffing Table: Payroll Cost for Maintenance Labor and Maintenance Supervision*

Position	Shift			Shifts/ day	Total per- sonnel	Wages, $/wh	Wages, $/day	Days/ yr	Annual cost, $	Salaried workers, $/yr	Total annual cost, $
	day	aft	night								
Machinist, B-4	2	1		3	4	22.76	182.08	365	199,400		199,400
Machinist, B-5	2	1		3	4	22.76	182.08	365	199,400		199,400
Machinist, B-6	2			2	3	22.76	182.08	365	133,000		133,000
Electrician, B-1	3	1	1	5	7	22.76	182.08	365	332,200		332,200
Electrician, B-2	1	1		2	3	22.76	182.08	365	133,000		133,000
Electrician, B-3	1	1		2	3	22.76	182.08	365	133,000		133,000
Electrician, B-4	1	1		2	3	22.76	182.08	365	133,000		133,000
Electrician, B-5	1	1		2	3	22.76	182.08	365	133,000		133,000
Electrician, B-6	1	1		2	2	22.76	182.08	260	94,600		94,600
Instrument mechanic, B-1	9	3	3	15	21	22.76	182.08	365	996,800		996,800
Instrument mechanic, B-2	2	1		3	4	22.76	182.08	365	199,400		199,400
Instrument mechanic, B-3	3	2	2	7	10	22.76	182.08	365	465,200		465,200
Instrument mechanic, B-4	3	2	2	7	10	22.76	182.08	365	465,200		465,200
Instrument mechanic, B-5	3	2	2	7	10	22.76	182.08	365	465,200		465,200
Instrument mechanic, B-6	1	1	1	3	4	22.76	182.08	365	199,400		199,400
Laborer or craft helper, all shops	60	25	21	106	148	19.16	153.28	365	5,930,400		5,930,400
General labor supervisor	4	2	2	8	11	19.36	154.88	365	452,200		452,200

Craft supervisor, services	2	1		3	4	24.04	192.32	365	210,600		210,600
Carpenter	5			5	5	21.36	170.88	260	222,200		222,200
Insulator/refractory	5			5	5	21.38	171.04	260	222,400		222,400
Painter	10			10	10	20.80	166.40	260	432,600		432,600
Crane operator	2	1		3	4	20.50	164.00	365	179,600		179,600
Truck driver	3	1	1	5	7	19.94	159.52	365	291,200		291,200
Sheetmetal worker	4			4	4	22.76	182.08	260	189,400		189,400
Mason	6			6	6	21.00	168.00	260	262,000		262,000
Utility repairer	3	2	2	7	10	22.76	182.08	365	465,200		465,200
Garage supervisor	1			1	1			260		64,000	64,000
Mechanic, garage	6	2	1	9	13	22.76	182.08	365	598,200		598,200
Secretarial pool	8			8	8			260		30,000	240,000
Mechanical engineer	1			1	1			260		72,000	72,000
Chemical engineer	1			1	1			260		72,000	72,000

* Includes fringe benefits.

mate. However, the actual depreciation which gives the owner the greatest tax advantage is actually used in the financial analysis.

If, indeed, the company has similar plants in operation, the operating costs should be checked against the historical data available. One should consider plant location (labor costs), utility rate, and freight rate differences when making comparisons.

Chapter 10 provides a more detailed discussion of the various components of operating costs and methods of calculating them.

QUESTIONS

9.1 What is the most general method used to determine labor for a preliminary estimate?

9.2 What information for the operating cost calculation is obtained from the buildup of the capital investment?

9.3 For a preliminary cost estimate, how is annual plant maintenance cost calculated?

9.4 What are the best two sources of information for determining the factor used to calculate plant maintenance?

9.5 What are the seven major elements of annual operating costs?

9.6 a) What is the range of percentage of payroll used to determine payroll overhead? b) What is the average percentage?

9.7 What items are included in plant operating supplies?

9.8 What operating expenses are included in indirect costs?

9.9 What are fixed costs related to?

9.10 To determine the method of depreciation, what must be considered?

9.11 Why are costs of salaried and shift workers treated differently in developing a staffing plan?

10

Estimating Operating and Manufacturing Costs

DEFINITIONS

To perform an operating or manufacturing cost estimate properly and to determine the potential profitability of a process, all costs must be considered in certain specific categories. The distinction between the various categories is quite important as they are treated differently for purposes of calculating taxes and profitability.

Direct costs are those which tend to be proportional or partially proportional to throughput or production. These costs include materials (raw materials, utilities, operating supplies, etc.) and labor (direct operating labor, operating supervision, maintenance supervision, payroll burden on labor charges, etc.).

Direct costs include both variable costs and semivariable costs. *Variable costs* are those such as raw materials and by-product credits which tend to be a direct function of production or plant output. In other words, these costs are maximized when a plant operates at full capacity and are not incurred when the plant is not operating.

Semivariable costs are direct costs which are only partially dependent upon production or plant output. These include such items as direct labor, supervision, maintenance, general expense, and plant overhead. While these costs increase with production rate, they do not increase in direct proportion to production rate. When the plant is not operating, a portion of the semivariable costs continues to be incurred. At zero production or throughput semivariable costs generally total about 20–40% of the total semivariable cost at full production.

Indirect costs, or *fixed costs*, are those costs which tend to be independent of production. These costs are incurred whether or not production rates change. They include plant overhead or burden (plant administration, laboratory, shops and repair facilities, property taxes, insurance, etc.). Indirect costs include depreciation.

Contingencies constitute an allowance which must be made in an operating cost estimate for unexpected or unpredictable costs or for error or variation likely to occur in the estimate. A contingency allowance is just as important in the operating cost estimate as it is in the capital cost estimate.

Another operating and manufacturing cost category is *distribution costs*. These are the costs associated with shipping the products to market. They include containers and packages, freight, operation of terminals and warehouses, etc.

General and administrative expenses, or *G&A expenses*, are those costs which are incurred above the factory or production level and which are associated with management. This category includes marketing and sales costs, salaries and expenses of officers and staff, accounting, central engineering, research and development, etc.

It is important in performing an operating cost estimate not to confuse general and administrative costs with *general works expense*. The latter item is a synonym for the fixed-cost factory overhead items, e.g., property taxes and insurance.

Further explanation of the above cost terms and many other cost engineering terms is provided in Appendix A.

TYPES OF OPERATING COST ESTIMATES AND ESTIMATING FORMS

As is true for a capital cost estimate, the purpose of an operating cost estimate is the controlling factor in determining the type of estimate to be performed. Preliminary or order-of-magnitude estimates are often used to screen projects and to eliminate uneconomical alternatives. More detailed estimates are then applied when the screening process has reduced the choice to a relatively few alternatives.

In performing the operating cost estimate, particularly on a preliminary basis, good judgment is necessary to avoid excessive attention to minor items which, even if severely over- or under-estimated, will not have a significant effect on the overall estimate. It is also necessary to calculate costs at reduced production rates as well as at design capacity. Operating costs are decidedly nonlinear with respect to production rate. This fact and the fact that virtually no plant or process operates all of the time at full design production rate make it imperative that reduced production rates be considered. This subject is discussed in considerable detail later in this chapter.

Finally, when estimating the effect of changes or additions to an existing process, the cost analysis should be performed on an incremental basis to evaluate the effect of the change as well as on an overall basis to determine if the entire project is worthy of being continued even without the change. Frequently a process change will not be economical, but the total project will be attractive. In other cases, the incremental costs of a change will appear to be quite profitable, but this profit will not be enough to offset losses entailed in the existing portion of the plant. Thus both types of analyses must be made.

Operating cost estimates can be performed on a daily, unit-of-production, or annual basis. Of these, the annual basis is preferred for the following reasons:

1. It "damps out" seasonal variations.
2. It considers equipment operating time.
3. It is readily adapted to less-than-full capacity operation.
4. It readily includes the effect of periodic large costs (scheduled maintenance, vacation shutdowns, catalyst changes, etc.).
5. It is directly usable in profitability analysis.
6. It is readily convertible to the other bases, daily cost and unit-of-production, yielding mean annual figures rather than a potentially high or low figure for an arbitrarily selected time of year.

As was emphasized in Chapter 2, a basic flowsheet of the process is vital to preparation of an estimate. This flowsheet should detail to the maximum extent possible the quantity, composition, temperature, and pressure of the input and output streams to each process unit.

In addition, to properly prepare an operating or manufacturing cost estimate, a prepared estimating form should be used to assure that the estimate is performed in a consistent manner and to avoid omitting major items. Figure 10.1 is an example of a suitable form for this purpose.

The estimating form acts as a checklist and as a device for cost recording and control. It must include the date of the estimate, the capital investment information which was previously determined, an appropriate cost index value reflecting the date of the capital cost estimate, the plant location, plant design capacity, annual anticipated plant operating days and/or annual production rate, and the plant or product identification. In any event, if a form is not used, the cost engineer should be equipped with a checklist and should be familiar with the technical aspects of the process. An estimate should never be made without specific technical knowledge of the process.

Last, wherever possible, cost data used in the estimate should be obtained from company records of similar or identical projects (with adjustment for inflation, plant site differences, and geography). For preliminary estimates, company records are probably the most accurate available source of cost data. If not available from company records, cost data may also be obtained from literature sources. Bear in mind, however, that such data are not always reliable for the reasons previously discussed in Chapter 2 on equipment pricing.

COST OF OPERATIONS AT LESS THAN FULL CAPACITY

The preceding discussion emphasized the necessity of performing operating and manufacturing cost estimates both at full plant capacity and at conditions other than full capacity. Frequently the inexperienced estimator will perform an estimate assuming operations only at full design capacity. This approach is totally erroneous as it does not consider unscheduled downtime, market fluctuations in product demand, time required to develop markets for a new product, and so forth.

Figure 10.2 is an illustration of cost effects of operation at less than full capacity. This figure takes into account the fixed, variable, and semivariable costs discussed earlier.

Semivariable costs, those which are partially proportional to production level, may include, among others, the following:

Direct labor
Supervision

General expense
Plant overhead

Other costs which may be semivariable, depending upon individual circumstances, are royalties and packaging. Packaging may be either variable or semivariable depending upon the particular situation. Royalties may be variable, semivariable, fixed, or even a capital expense. Thus they must be carefully examined to be certain that they are included in the proper cost category. A royalty fee which is paid in a lump sum should be capitalized. Royalties which are paid in equal annual increments are treated as fixed costs. Those paid as a fee per unit of production or sales are variable costs, and those which are paid in a sliding scale, i.e., at a rate per unit of production which declines as production increases, are semivariable. In certain cases royalty agreements may contain elements of more than one cost category—for example, an annual fee (fixed) plus a charge per unit of production (variable).

Fixed-cost items, in addition to royalties if applicable, include the following:

Depreciation
Property taxes
Insurance

Variable costs generally include the following:

Raw materials
Utilities
Royalties (if applicable)
Packaging (if applicable)
Marketing
Catalysts and chemicals

Figure 10.2 graphically demonstrates the implications of operating at less than full capacity. In this figure, at 100% of capacity, the following apply:

F = the fixed expense

V = the variable expense

Date ______

By ______

PRELIMINARY PRODUCTION COST ESTIMATING FORM

Location: ______		Product(s): ______			
Capital Investment:		Process: ______			
Total	______	Nelson Index:	______	CE Index	______
Less working capital	______	M&S Index	______	Annual Operating	
Less salvage value	______	ENR Index:	______	Days:	______
Depreciable investment	______	Annual production: ______			

	Raw materials	Annual quantity	Unit cost	\$/year	\$/
(1)	______	______	______	______	______
(2)	______	______	______	______	______
(3)	______	______	______	______	______
(4)	______	______	______	______	______
(5)		Gross raw material cost (sum of lines 1 to 4):		______	______
	Misc. credits and debits				
(6)	______	______	______	______	______
(7)	______	______	______	______	______
(8)	______	______	______	______	______
(9)		Total debit (credit) (sum of lines 6 to 8):		______	______
(10)		Net raw material cost (lines 5 + line 9):		______	______

Figure 10.1a Preliminary estimating form, part I.

Direct expense	Unit	Quantity	Unit cost	$/year	$/
(11) Steam	M lb	______	______	______	______
(12) Water (　　)	M gal	______	______	______	______
(13) Water (　　)	M gal	______	______	______	______
(14) Electricity	kW-hr	______	______	______	______
(15) Fuel (　　)	______	______	______	______	______
(16) Fuel (　　)	______	______	______	______	______
(17) Labor	______			______	______
(18) Supervision	______			______	______
(19) Maintenance	______			______	______
(20) Factory supplies	______			______	______
(21) Indirect overhead	______			______	______
(22) Payroll overhead	______			______	______
(23) Laboratory	______			______	______
(24) Contingencies	______			______	______
(25)	Total direct conversion cost (sum of lines 11 to 24):			______	______
Indirect expense					
(26) Depreciation	______			______	______
(27) Real estate taxes & insurance	______			______	______
(28) Depletion allowances	______			______	______
(29) Amortization	______			______	______
(30)	Total indirect conversion cost (sum of lines 26 to 29):			______	______
(31)	Total conversion cost (line 25 + line 30):			______	______
(32)	Total operating cost (line 31 + line 10):			______	______
(33) Packing & shipping expense	______			______	______
(34)	TOTAL COST FOR PLANT (line 32 + line 33):			______	______

Figure 10.1b Preliminary estimating form, part II.

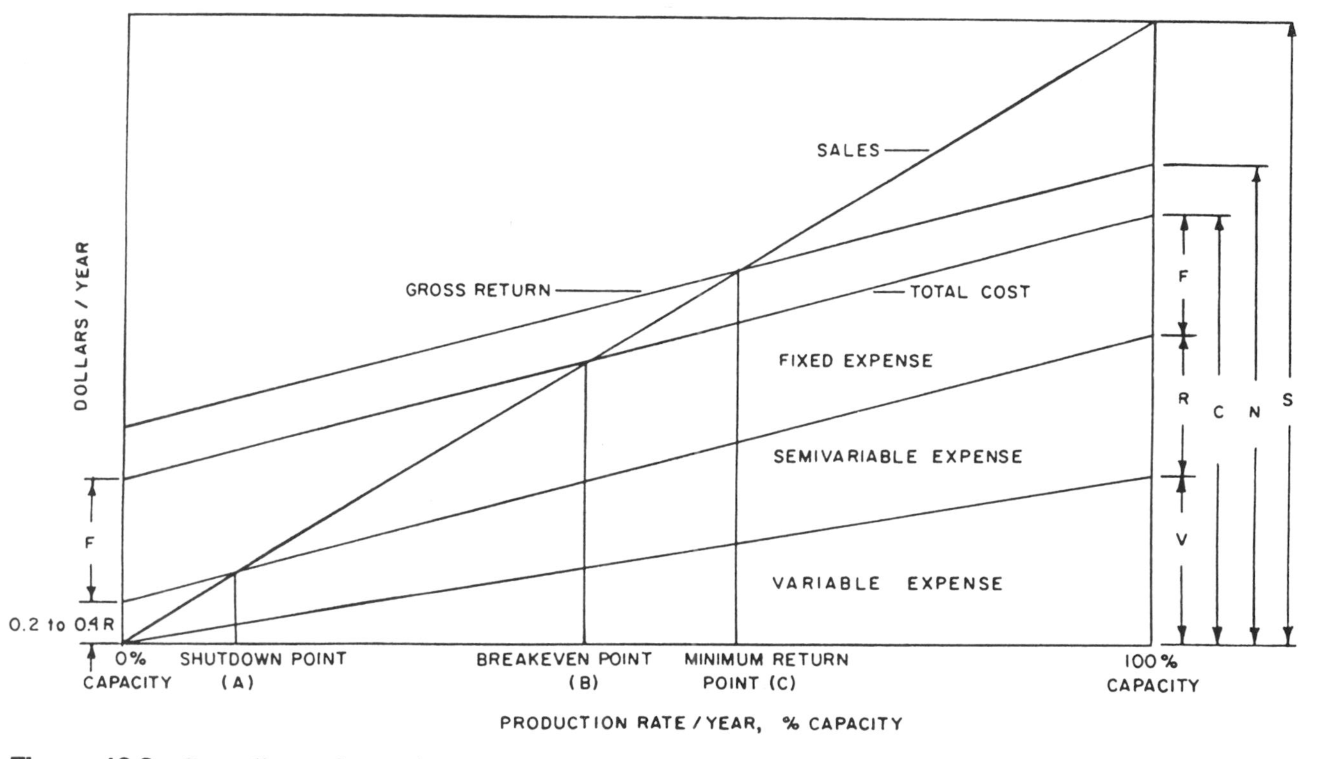

Figure 10.2 Cost effects of operations at less than full plant capacity.

R = the semivariable expense

C = total operating cost

S = sales income

N = the income required to achieve the minimum acceptable return on investment before taxes (P) for the capital investment I

As can be seen from the figure, the variable expense declines to zero at 0% of capacity, fixed expense is constant, and semivariable expense declines at 0% of capacity to from 20–40% of its value at full capacity.

This simple plot is used to determine the following:

The minimum production rate at which the desired return on investment will be achieved (C)
The breakeven point, that point at which income will exactly equal total operating cost (B)
The shutdown point, that point at which it is advisable to shut down the plant rather than operate at lower production rates (A)

The plot readily identifies the range of production rates at which the following apply:

The return on investment will equal or exceed the desired minimum (all production rates $\geq C$).
The return on investment will be less than the desired value but will be greater than zero (production rates $< C$ but $> B$).
The process will result in a loss, but losses will be minimized by continuing to operate the plant rather than shutting it down (production rates $\leq B$ but $> A$).
Losses will become so large that it is less expensive to close the plant and pay fixed expenses out of pocket rather than to continue operations (production rates $\leq A$).

The breakeven and shutdown points can also be determined mathematically as follows:

$$B \text{ (breakeven point)} = \frac{(F + nR)}{S - V - (1 - n)R}$$

where

n = decimal fraction of semivariable costs incurred at zero production (usually about 0.3)

$$A \text{ (shutdown point)} = \frac{nR}{S - V - (1-n)R}$$

Similarly, the total cost line can be expressed as

$$C_p = [V + (1-n)R]p + F + nR$$

where

C_p = total cost at production rate p

p = actual annual production rate as a fraction of plant capacity

Since total annual sales are proportional to production (assuming no stock-piling of production), and therefore have no value at zero output, the equation for the sales line is

$$S_p = (S \times p)$$

where

S_p = sales income at production rate p.

The point at which the sales and total cost lines cross is the breakeven point for the plant and is equal to the level of output at which sales is equal to total cost.

RAW MATERIALS COSTS

Depending on the particular process, raw materials costs can constitute a major portion of operating costs. For this reason, a complete list of all raw materials must be developed using the process flowsheet as a guide. In developing the raw materials list, the following information must be obtained for each raw material:

Units of purchase (tons, pounds, etc.)
Unit cost
Available sources of the material
Quantity required per unit of time and/or unit of production
Quality of raw materials (concentration, acceptable impurity levels, etc.)

In estimating the quantity of each raw material, appropriate allowance must be made for losses in handling and storage, process waste, and process yield.

Price data for purchased raw materials are generally available to a high level of accuracy from many sources. Purchased price information can generally be obtained from the suppliers. Alternatively, supplier catalogs and price lists can be used, as can published data. The *Wall Street Journal, Iron Age, Oil Paint and Drug Reporter, European Chemical News*, and similar trade and business publications are all good sources of spot data.

In estimating the cost of any raw material it must be remembered that in general raw material costs vary with quality (concentration, surface finish of metals, impurities, etc.) and generally decrease in unit cost as quantity increases. There is little sense, for example, in relying on a cost figure for reagent-grade hydrochloric acid in 5–lb bottles when the process can utilize much-lower-cost commercial-grade hydrochloric acid (muriatic) in tank car or tank truck lots.

Another major factor to be considered is availability of the raw material. Does sufficient productive capacity exist such that the market can supply the demands of the proposed process? A sudden large new demand for a raw material can, and often does, cause substantial price increases, particularly if the material is available only in small quantities or as a by-product of another process.

In pricing raw materials it must also be remembered that the prices are generally negotiated and discounts obtained can result in prices substantially less than quoted or published data. Where available, company experience in negotiating supply orders for the same or similar materials should be used to estimate the amount of any such probable discounts.

A common pitfall in operating and manufacturing cost estimates is to neglect the cost of raw materials manufactured or obtained in-house or from another company division because they are not purchased. Such raw materials, however, do represent a cost to the company and do have value. In the estimate, therefore, they should be included as a cost at their market value or company book value. The market value to be used is the going market price corrected for any direct sales costs which are not incurred due to internal use. In addition, internal company freight, handling, and transfer costs must be added.

If the captive raw material is an intermediate product which has no established market price, the cost should be based upon the value of the nearest downstream product or material for which an established market price exists. The cost is equal to the value of the downstream product less direct sales costs not incurred plus internal company transfer costs less manufacturing costs for operations avoided by using the intermediate product instead of processing it further.

Other raw materials cost items which are easily overlooked are fuels which are used as raw materials (e.g., natural gas in methane conversion processes) and the periodic makeup of losses to catalysts and other processing materials. Fuels which are used as a raw material should be treated in the same manner as any other raw material. Those which are used for utility purposes should be treated as a utility cost. In the case of catalysts and similar materials, the initial fill of such items is usually treated as a capital expense if its useful life exceeds one year. Otherwise it is considered as a start-up expense.

Finally, in estimating raw materials costs it must be recognized that prices are usually quoted FOB the supplier's plant or basing point, not at the point of use. Thus freight to the point of use and local handling costs must be added to the quoted prices. Current freight tariffs for each commodity should be checked in preparing the estimate since they are complex and often illogical and cannot be generalized.

BY-PRODUCT CREDITS AND DEBITS

All process by-products, including wastes and pollutants, must be considered in the operating cost estimate. Thus, every output stream shown on the process flowsheet must have a cost assigned to it. Obviously, these costs may be credits (in the case of salable or usable by-products) or debits (in the case of wastes or unsalable by-products).

Items which are not immediately obvious as by-products are nuisance expenses. No longer can a plant discharge pollution at will into the air or water nor can high noise levels, thermal discharges, odors, etc., remain unabated. These are all by-products, albeit undesirable ones, and the cost of treating them must be included in the estimate—either in the overall process costs or as a by-product debit. Similarly, the capital cost estimate must include the costs of associated pollution and nuisance abatement equipment.

Nuisance costs are also not confined to things which leave or are discharged from the plant. In many cases, nuisances which are entirely confined to the plant premises, e.g., high noise levels, must be eliminated due to safety and employer liability considerations.

Salable by-products have values which can be established in the same manner as determining the cost of raw materials. The by-product credit then may be estimated from the market prices or anticipated selling prices of the by-products less any costs of processing, packaging, selling and transporting them to market. In many cases it is necessary to do a complete capital and operating cost estimate on the by-product processing facilities in order to determine the latter costs.

If the by-product is not sold but instead is converted to another salable product, it is valued at the market value of the subsequent product less the associated conversion and transfer costs. Wastes similarly carry the negative value associated with their treatment and disposal. Again, it may be necessary to perform another complete capital and operating cost estimate to determine these costs.

Credits should be taken for by-products with great care. Many processes have been economic failures because assumed by-product credits were not realized. In many cases, the production of a by-product can glut its market, particularly if the current market is a small one. Similarly, introduction of the by-product into competition with other materials can depress prices due to competitive pressures.

A notable example of this type of situation is the sulfur market. In the mid-1960s sulfur was in short supply and carried an inflated price. At that time, many estimates were being made on pollution-control devices and systems which would produce sulfur as a by-product and were justified on the basis of the high by-product credits for sulfur. However, due to improved methods of sulfur production and the additional sulfur being produced as a by-product, in a very short period of time the market for sulfur and sulfur products (e.g., sulfuric acid) changed to a state of oversupply and depressed prices. As a result, many supposedly profitable processes actually became uneconomical.

UTILITY COSTS

The estimation of utility costs, particularly in light of rapidly increasing energy costs, is another critical area of operating cost estimation.

In estimating utility costs, it is necessary first to determine the requirements for each utility including a reasonable allowance for nonproduction items such as plant lighting, sanitary water, etc. An allowance should also

Table 10.1 Typical Utility Summary

	Power (kWh/hr)	Water required (gpm)	Water recovery and makeup (gpm)
Mine	1,830	1,000	
Crushing plant	1,920		
Concentrating plant	18,550	38,650	36,720
Pelletizing plant	4,050	1,330	
Subtotal:	26,350	40,980	36,720
Utilities:			
Makeup water	400		5,580
Plant lighting	300		
Sanitary water	150	500	
General facilities	150	300	
Miscellaneous and contingencies	150	520	
Total:	27,500	42,300	42,300

be made for miscellaneous usage and contingencies. Table 10.1 is a typical utility summary of the nature that is required.

In addition to the utility summary, consumption patterns should be examined to determine if consumption will be at a uniform rate during each day and from day to day. If consumption rates fluctuate, utility pricing may be based not only on total consumption but also in the peak demand rate and, in some cases, the time of day in which the peak occurs. A further consideration is whether or not the utility must be available on a noninterruptable basis. If the process has an alternate utility source or can tolerate periodic cutbacks in utility supplies, rates may be available which are somewhat lower in cost than noninterruptable rates. The estimate should also consider the fact that in certain areas, noninterruptable rates are not available and electric power, natural gas, etc., can be curtailed by the utility companies at any time if residential demands exceed available supplies. In these situations both the capital and operating cost estimates must consider standby alternative utility sources such as emergency power generators, combustion units which can operate on alternate fuels, etc.

In general, utility costs decrease as demand increases (although there is a growing tendency toward national and state policies to increase unit cost with quantity artificially in order to encourage reductions in consumption).

Electric power charges are usually based upon a demand factor—the maximum power draw during a 15- to 30-min period in any given month. A load factor is computed as the ratio of average usage to the demand factor and rates are established to give preference to high load factors (i.e., steady consumption). The rate schedules also generally include an escalation factor which is tied to increases in fuel costs.

Electric rates in the past were remarkably stable for many, many years, and it was common estimating practice to arbitrarily assume a cost of the order of a cent or two per kW-hr for preliminary estimating purposes. This is no longer true, and no generalization can be made about electric rates. The estimator must obtain current rates from the utility companies serving the proposed plant and can no longer safely assume any "rule-of-thumb" figure for power costs. If the power is not purchased and instead is obtained from a captive company-owned generating system, utility costs must be based upon a study of the system itself.

Natural gas prices depend on quantity required. Steam costs are dependent upon many factors including pressure, cost of fuel, temperature, credit for heating value of condensate, etc. If company data are not available on steam cost, it must be estimated taking into account fuel cost, boiler water treatment, operating labor, depreciation on investment, maintenance and other related costs of steam production. Black (1) has suggested that steam costs can be approximated as 2–3 times the cost of fuel.

Water costs are highly variable depending upon the water quality needed and the quantity required. Purification costs, if contamination occurs before disposal, must also be included, as must cooling costs if the process results in heating of process water. In most jurisdictions, water may not be discharged into streams or the natural water table unless it is equal or better in quality and temperature as when it was withdrawn from the stream.

Fuel costs vary with the type of fuel used, the Btu value of the fuel, and the source of supply. Careful consideration should be given in the estimate not only to fuel cost but to the type of firing equipment required and to required fuel storage facilities. Often these factors can rule out what would otherwise be the least expensive available fuel.

Another factor to be considered is that, as mentioned earlier in this chapter, certain fuels, although lower in cost than alternative fuels, may not be available in sufficient quantities, if at all. In some cases, a fuel may be abundant in warm weather but be in short supply during the heating season, necessitating use of alternate fuel supplies or planned production cutbacks or stoppages during periods of severe winter weather.

In estimating utility requirements, equipment, efficiency losses, and contingencies must be considered. Utility consumption generally is not proportional to production due to economies of scale and reduced energy losses per unit of volume or production on larger process units. Black (1) has suggested that utility consumption varies to the 0.9 power of capacity, rather than in direct proportion to capacity.

Last, while not normally thought of as a utility, cost of motor fuels and greases for all mobile equipment must be estimated. Fuel costs are based upon annual operating hours for each piece of equipment times fuel consumption per hour times the prevailing cost for the fuel to be used. Greases and lubricant costs are directly related to fuel costs, and for most types of equipment amount to approximately 16% of the fuel cost.

LABOR COSTS

Labor costs, particularly in a labor-intensive process, may be the dominant cost factor in an operating or manufacturing cost estimate. To properly estimate these costs, a staffing table must be established in as detailed a manner as possible. This table should indicate the following:

1. The particular skill or craft required in each operation
2. Labor rates for the various types of operations
3. Supervision required for each process step
4. Overhead personnel required

It is not always possible to determine the extent of supervision and overhead personnel which are required. In such cases, alternate methods of estimating these costs may be used as discussed later in this book. However, if sufficient data are available, these factors should be included in the staffing table for maximum estimate accuracy.

Table 10.2 is a typical staffing table for a complete estimate. This table illustrates the detail which is required in a complete staffing table. Note that the table includes general and administrative personnel, production workers, maintenance workers, and direct supervision.

Once the staffing table is developed (at a minimum including all direct production labor), labor costs can readily be estimated from company records of wages and salaries by position, union wage scales, salary surveys of various crafts and professions, or other published sources. Because labor rates are prone to rapid inflation, often at rates sharply different from general inflation rates, care must be taken to obtain current figures and to properly project future wage rates.

Generally, data on wage rates includes shift differentials and overtime premiums. If not, these factors must be added to the extent applicable. Further, when estimating around-the-clock, 168–hr/wk operations, allowance must be made for the fact that a week includes 4.2 standard 40–hr weeks. Even with four work crews on "swing shift," one crew must work 8 hr/week of overtime to keep the plant in steady operation. Depending upon local custom, laws, and union contracts, this overtime is generally payable at 1.5–3 times the normal hourly rate.

An alternate method of calculating labor requirements, if sufficient data are not available to establish a staffing table, is to consider a correlation of labor in workhours per ton of product per processing step. This relationship, which was developed by Wessell (2), relates labor requirements to plant capacity by the following equation:

$$\frac{\text{Operating workhours}}{\text{tons of product}} = t\left[\frac{\text{number of processing steps}}{(\text{capacity, tons/day})^{0.76}}\right]$$

where

t = 23 for batch operations with a maximum of labor

t = 17 for operations with average labor requirements

t = 10 for well-instrumented continuous process operations

As pointed out previously, the relationship between labor requirements and production rate is not usually a direct one. The Wessell equation recognizes that labor productivity generally improves as plant throughput increases. It can also be used to extrapolate known workhour requirements from one plant to another of different capacity.

Another shortcut method of estimating labor requirements, when requirements at one capacity are known, is to project labor requirements for other capacities to the 0.2–0.25 power of the capacity ratio.

Table 10.2 Typical Complete Staffing Table

	Shifts/ day	Days/ year	Number of labor personnel	Wages for labor			Number of salaried personnel	Salaries		Total wages & salaries ($)
				Hour	Day	Year		per person	Total	
General										
Administration and services										
General superintendent	1						1	80,000	80,000	80,000
Chief engineer	1						1	60,000	60,000	60,000
Personnel officer	1						1	36,000	36,000	36,000
Personnel clerk	2						2	24,000	48,000	48,000
Payroll clerk	3						3	30,000	120,000	120,000
Accountant	3						3	36,000	108,000	108,000
Guard	6	365	8.4	11.04	529.92	193,600				193,600
Stock clerk	6	365	8.4	11.04	529.92	193,600				193,600
Safety engineer	1						1	48,000	48,000	48,000
Chief chemist	1						1	52,000	52,000	52,000
Drafter	3						4.2	28,000	117,600	117,600
Sample collector & tester	12						16.8	36,000	604,800	604,800
Stenographer	3						3	24,000	72,000	72,000
Subtotal:						387,200			1,316,400	1,703,600

Mine										
Administration and Services										
Mine superintendent	1						1	72,000	72,000	72,000
Mining engineer	2						2	56,000	112,000	112,000
Geologist	1						1	48,000	48,000	48,000
General foreman/ forewoman	1						1	52,000	52,000	52,000
Rod carrier	2	260	2	12.04	192.64	50,000				50,000
Clerk	2						2	28,000	56,000	56,000
Warehouser	3	260	3	13.04	312.96	81,200				81,200
Janitor	2	260	2	11.04	176.64	46,000				46,000
Party chief	1						1	48,000	48,000	48,000
Subtotal:						177,200			388,000	565,200
Production										
Shovel operator	9	260	9	18.36	1321.92	343,600				343,600
Shovel oiler	6	260	6	12.04	577.92	150,400				150,400
Truck driver, haulage	24	260	24	14.04	2695.68	700,800				700,800
Powderer	1	260	1	13.72	109.76	28,400				28,400
Powderer helper	2	260	2	12.04	192.64	50,000				50,000
Truck driver: service truck, fuel truck	6	260	6	13.04	625.92	162,800				162,800
Operator, crane	3	260	3	13.04	312.96	81,200				81,200
Operator, drill	9	260	9	17.04	1226.88	318,800				318,800
Operator helper, drill	9	260	9	12.04	866.88	225,200				225,200

Table 10.2 *(Continued)*

	Shifts/ day	Days/ year	Number of labor personnel	Wages for labor Hour	Day	Year	Number of salaried personnel	Salaries per person	Total	Total wages & salaries ($)
Operator, tractor	9	260	9	13.40	964.80	250,800				250,800
Laborer	18	260	18	11.72	1687.68	438,800				438,800
Shift foreman/ forewoman	3						3	52,000	156,000	156,000
Dumper	3	260	3	11.40	273.60	71,200				71,200
Front end loader operator	1	260	1	13.72	109.76	28,400				28,400
Primary crusher operator	3	260	3	13.56	325.44	84,400				84,400
Primary crusher general laborer	4	260	4	11.04	353.28	91,600				91,600
Subtotal:						3,026,400				3,026,400
Maintenance										
Truck mechanic	10	260	10	17.40	1536.00	362,000				362,000
Shovel mechanic	3	260	3	17.40	417.60	108,400				108,400
Electrician	4	260	4	18.04	577.28	150,000				150,000
Tire repairer	3	260	3	16.68	400.32	104,000				104,000
Laborer	9	260	9	11.04	794.88	206,800				206,800

Welder	5	260	5	17.40	696.00	180,800				180,800
Shift foreman/ forewoman	4						4	52,000	208,000	208,000
Chief electrician	1						1	52,000	52,000	52,000
Carpenter	3	260	3	17.04	408.96	106,400				106,400
Master mechanic	1	260	1	18.04	144.32	37,600				37,600
Shop foreman/ forewoman	1						1	52,000	52,000	52,000
Machinist	1	260	1	18.04	144.32	37,600				37,600
Painter	1	260	1	14.04	112.32	29,200				29,200
Subtotal:						1,322,800			312,000	1,634,800
Plant										
Administration and services										
Plant manager	1						1	64,000	64,000	64,000
Mechanical engineer	2						2	48,000	96,000	96,000
Metallurgist	2						2	48,000	96,000	96,000
Chemist	4						4	44,000	176,000	176,000
Stenographer	2						2	24,000	48,000	48,000
Clerk	2						2	24,000	48,000	48,000
Warehouser	4	365	5.6	13.04	417.28	152,400				152,400
Electrical engineer	1						1	48,000	48,000	48,000
Sampler	6	365	8.4	12.72	610.56	222,800				
Subtotal:						375,200			576,000	951,200

Table 10.2 *(Continued)*

	Shifts/ day	Days/ year	Number of labor personnel	Wages for labor			Number of salaried personnel	Salaries		Total wages & salaries ($)
				Hour	Day	Year		per person	Total	
Production										
General foreman/ forewoman	1						1	60,000	60,000	60,000
Shift foreman/forewoman, crushing and screening	3						4.2	52,000	218,400	218,400
Shift foreman/fore-woman, concen-trating	3						4.2	52,000	218,400	218,400
Shift foreman/fore-woman, pelletizing	3						4.2	52,000	218,400	218,400
Labor foreman/ forewoman	1						1	52,000	52,000	52,000
Fine crusher panel operator	3	365	4.2	13.56	325.44	118,800				118,800
Fine crusher operator, roving	6	365	8.4	13.56	650.88	237,600				237,600

Crushing, general labor	8	365	11.2	11.04	706.56	258,000		258,000
Concentrator panel operator	3	365	4.2	13.56	325.44	118,800		118,800
Concentrator assistant panel operator	3	365	4.2	12.04	288.96	105,600		105,600
Grinding floor operator, roving	3	365	4.2	16.68	400.32	146,000		146,000
Filter operator, roving	3	365	4.2	12.72	305.28	111,600		111,600
Pump thickener operator, roving	3	365	4.2	12.04	288.96	105,600		105,600
Concentrator, general labor	10	365	14.0	11.04	883.20	322,400		322,400
Grate kiln operator	3	365	4.2	17.04	408.96	149,200		149,200
Grate kiln assistant operator	3	365	4.2	13.40	321.60	117,200		117,200
Grate kiln operator, roving	3	365	4.2	17.04	408.96	149,200		149,200
Balling drum operator, roving	3	365	4.2	13.40	321.60	117,200		117,200
Pellet load-out operator	3	365	4.2	13.40	321.60	117,200		117,200
Pellet load-out loader	3	365	4.2	12.04	288.96	105,600		105,600
Pelletizing-general laborer	10	365	14	11.04	883.20	322,400		322,400
Front end load operator	1	260	1	13.04	104.32	27,200		27,200
Grader operator	1	260	1	13.40	107.20	28,000		28,000
Tractor operator	1	260	1	13.40	107.20	28,000		28,000

Table 10.2 (*Continued*)

	Shifts/ day	Days/ year	Number of labor personnel	Wages for labor			Number of salaried personnel	Salaries		Total wages & salaries ($)
				Hour	Day	Year		per person	Total	
Boom truck operator	1	260	1	13.04	104.32	27,200				27,200
Subtotal:						2,712,800			767,200	3,480,000
Maintenance										
General foreman/ forewoman	1						1	60,000	60,000	60,000
Foreman/forewoman, concentrator	2						2	52,000	104,000	104,000
Foreman/forewoman, crusher	1						1	52,000	52,000	52,000

Foreman/forewoman, pelletizer	2						2	52,000	104,000	104,000
Millwright, crusher	6	365	8.4	17.40	835.20	304,800				304,800
Millwright, concentrator	10	365	14	17.40	1392.00	508,000				508,000
Millwright, pelletizer	10	365	14	17.40	1392.00	508,000				508,000
Oilers, concentrator	2	365	2.8	12.04	192.64	70,400				70,400
Oilers, crusher	1	260	1	12.04	96.32	25,200				25,200
Oilers, pelletizer	1	365	1.4	12.04	96.32	35,200				35,200
Welders	4	260	4	17.40	556.80	144,800				144,800
Machinists	2	260	2	18.04	288.64	75,200				75,200
Blacksmiths	2	260	2	17.72	283.52	73,600				73,600
Tool room attendant	2	260	2	11.72	187.52	48,800				48,800
Foreman/forewoman, electrical	2						2.8	52,000	145,600	145,600
Electrician	12	365	16.8	18.04	1731.84	632,000				632,000
Subtotal:						2,426,000			465,600	
Total:						10,427,600			3,981,200	14,408,800

A significant factor to be considered in estimating labor costs is overtime. As mentioned earlier, around-the-clock, 24–hr/day, 7–day/week operations have an inherent overtime penalty of 8 hr/week. With this exception, and occasional overtime to cover for absent workers, overtime is usually not a major consideration in manufacturing and production operations. However, in estimates involving construction projects, or those which anticipate regular scheduled overtime, these costs can be substantial and must be carefully evaluated.

Scheduled overtime over an extended period can result in substantial decreases in worker productivity resulting in a major cost penalty for productivity losses in addition to the higher direct costs of premium pay at 1.5, 2, or even 3 times normal hourly rates.

Scheduled overtime involves a planned, continuing schedule for extended working hours for individual workers or even entire crews. It is not occasional overtime caused on an irregular basis by absenteeism, equipment malfunctions, etc.

Unfortunately, scheduled overtime rarely saves money or accelerates production. Two articles which appeared in the *AACE Bulletin* in 1973 (3,4) amply illustrated this point. These articles, prepared by representatives of the Construction Users Anti-Inflation Roundtable (now Business Roundtable), clearly demonstrated that scheduled overtime rarely, if ever, is beneficial, and that overtime should be avoided in favor of additional employees working normal shifts or partial shifts. Quoting from one article:

> Studies indicate that when a job is placed on overtime there is a sharp drop in productivity during the first week with a substantial recovery which holds for about two weeks and is followed by a fairly steady decline. At the end of seven to nine weeks the productivity on an overtime basis is no greater than the productivity would be on a 40–hour week. In this period there is an increase in work accomplished of about 12%. After seven to nine weeks of operation, productivity continues to decline and the work accomplished is less than would have been accomplished on a 40 hour per week schedule. After 18 to 20 weeks there is no gain in total work accomplished.... [see Figure 10.3.]
>
> A job requiring 24 weeks to complete at 40 hours per week is shown on this chart ... [see Figure 10.4]. If we want to complete the job in 16 weeks, a mathematical calculation of the hours of work involved would indicate a 60–hour week would accomplish it.

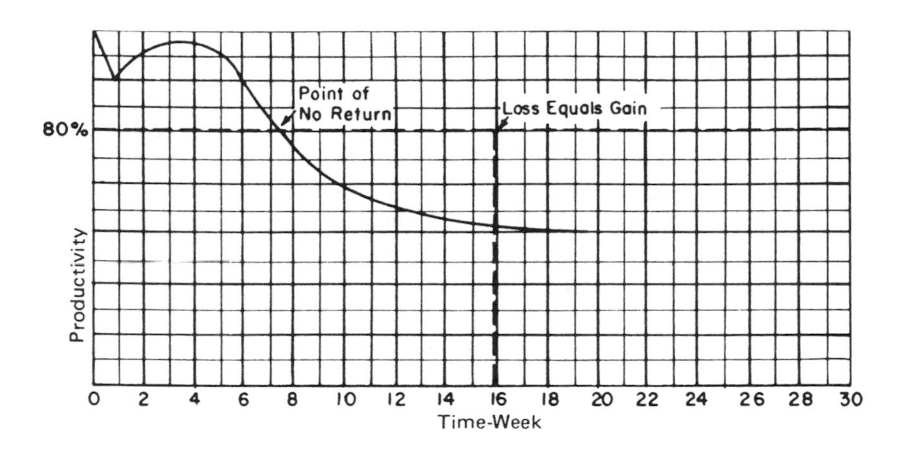

Figure 10.3 Effect of scheduled overtime on productivity.

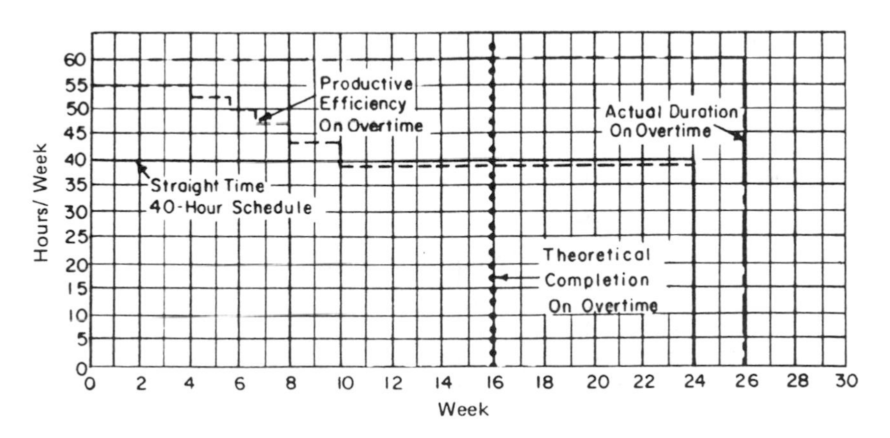

Figure 10.4 Effect of scheduled overtime on job completion time.

> Assuming we maintain the same number of workers, actual accomplishment of work indicates the job would require 26 weeks due to the reduced productivity. This means we would actually lose two weeks in schedule time even though the mathematics indicate we could improve the schedule by 8 weeks. (4)

The impact of the above situation at overtime rates of double regular rates is that the 60–hr/week scheduled overtime job results in a labor cost of 217% of the regular 40–hr cost due to productivity losses. Thus scheduled overtime is clearly no bargain.

The magnitude of the problem of scheduled overtime is further shown in the other article (3) as illustrated in Figures 10.5 through 10.8. Clearly there is little sense in planning a project from the onset on a basis of scheduled overtime. Thus it is imperative that the cost engineer who is responsible for performing a cost estimate uses good judgment and recommends changes in labor and staffing tables to avoid scheduled overtime.

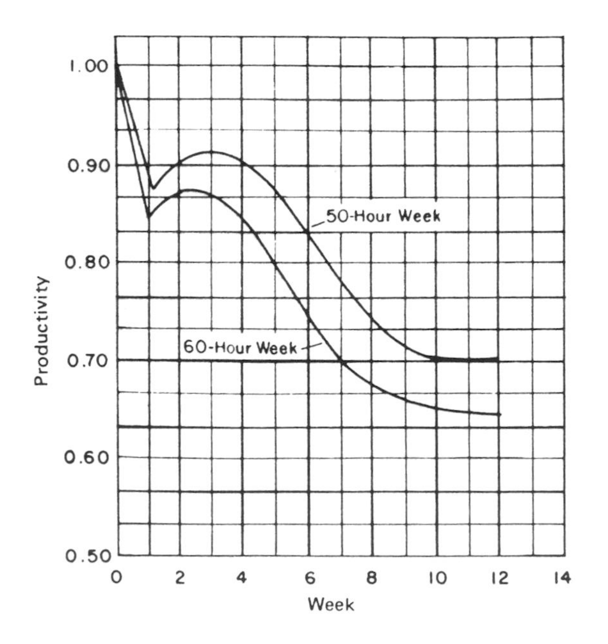

Figure 10.5 Cumulative effect of overtime of productivity: 50–and 60–hr work weeks.

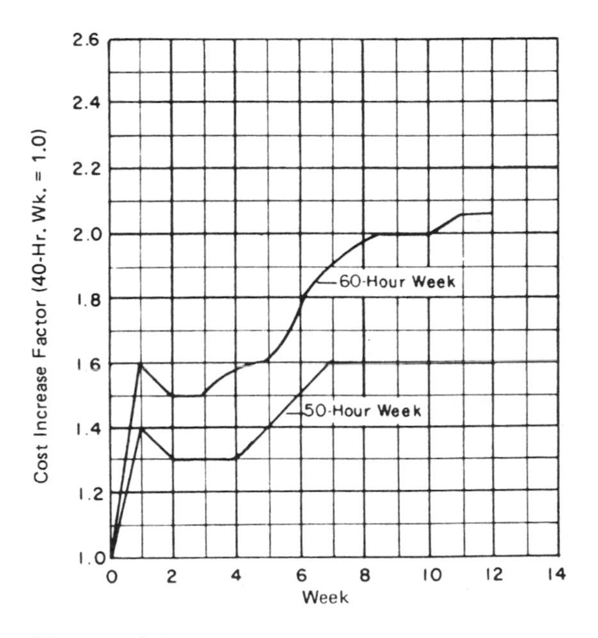

Figure 10.6 Cumulative cost of overtime: successive weeks productivity loss plus overtime premiums.

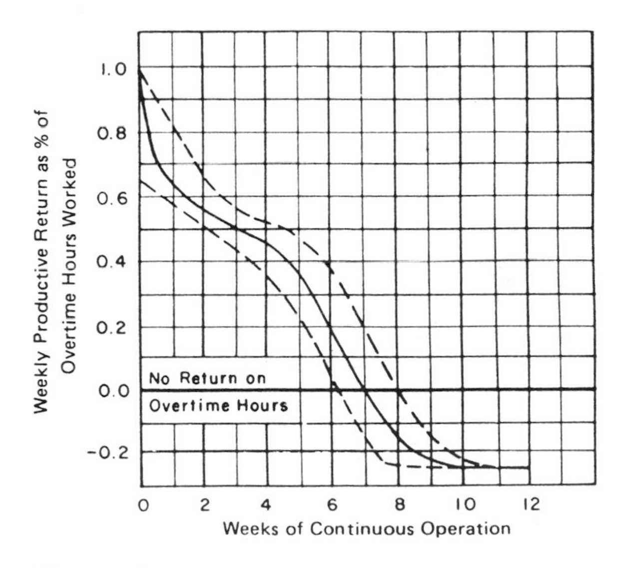

Figure 10.7 Percentage of effective return on overtime hours for 50–hr job schedule.

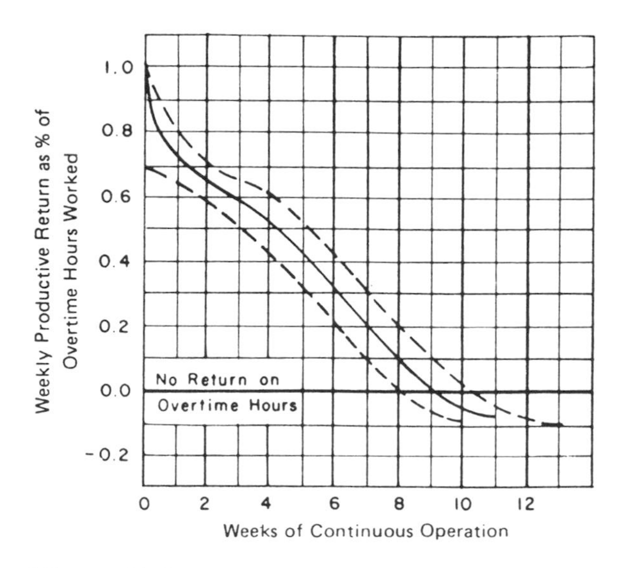

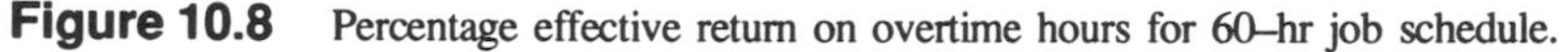
Figure 10.8 Percentage effective return on overtime hours for 60–hr job schedule.

SUPERVISION AND MAINTENANCE COSTS

Supervision costs should be established, if at all possible, through a staffing table and tabulation of associated costs. Unfortunately, for most preliminary estimates, particularly those for proposed new processes, this is not possible. In such cases, costs of supervision can be roughly estimated by taking a fixed percentage of direct labor costs based upon company experience. In the absence of prior data on similar operations, a factor of 15–20% is generally satisfactory.

The validity of the latter factor can readily be seen when considering the fact that one front-line supervisor can effectively manage no more than 8–10 workers, i.e., supervision workhours of 0.100–0.125 per direct labor workhour. With frontline supervision (i.e., foremen or forewomen) at labor rates approximately 50–60% above general labor rates, a 15–20% supervision factor is evident.

Maintenance labor costs, like supervision costs, should be delineated in the labor staffing table if at all possible. However, other than company records of existing similar plants, reliable data on maintenance costs are generally not available, and the staffing table approach is usually not feasible.

For this reason, maintenance costs are often estimated as a fixed percentage of depreciable capital investment per year. For complex plants and severe corrosive conditions, this factor can be 10–12% or higher. For simple plants with relatively mild, noncorrosive conditions, 3–5% should be adequate.

Maintenance costs are a semivariable category which is generally distributed about 50% to labor and 50% of materials. For a preliminary estimate the various factors making up plant maintenance can be back calculated from the total maintenance number using the following approximate percentages:

Direct maintenance labor, 35–40%
Direct maintenance labor, supervision, 7–8%
Maintenance materials, 35–40%
Contract maintenance, 18–20%

Then as the project evolves toward a final staffing plan, the factors can be improved and are finally replaced with numbers generated from the staffing table. The back calculation allows one to estimate the number of people required early in the project.

When operating at less than 100% of capacity, maintenance costs generally increase per unit of production. Such operations can be estimated as follows:

% of capacity	Maintenance cost as a percentage of cost at 100% capacity
100	100
75	85
50	75
0	30

Maintenance generally increases with age of equipment although most estimates use an average figure to simplify the estimate. This apparent error is offset in the overall estimate by use of average or "straight-line" depreciation, whereas accelerated depreciation is in fact generally used for tax purposes. Figure 10.9 illustrates the validity of using an average maintenance figure over the life of a project.

Finally, for major projects, it may be necessary to include costs for additional maintenance supervisors. However, for small plant additions, additional supervision is generally not required.

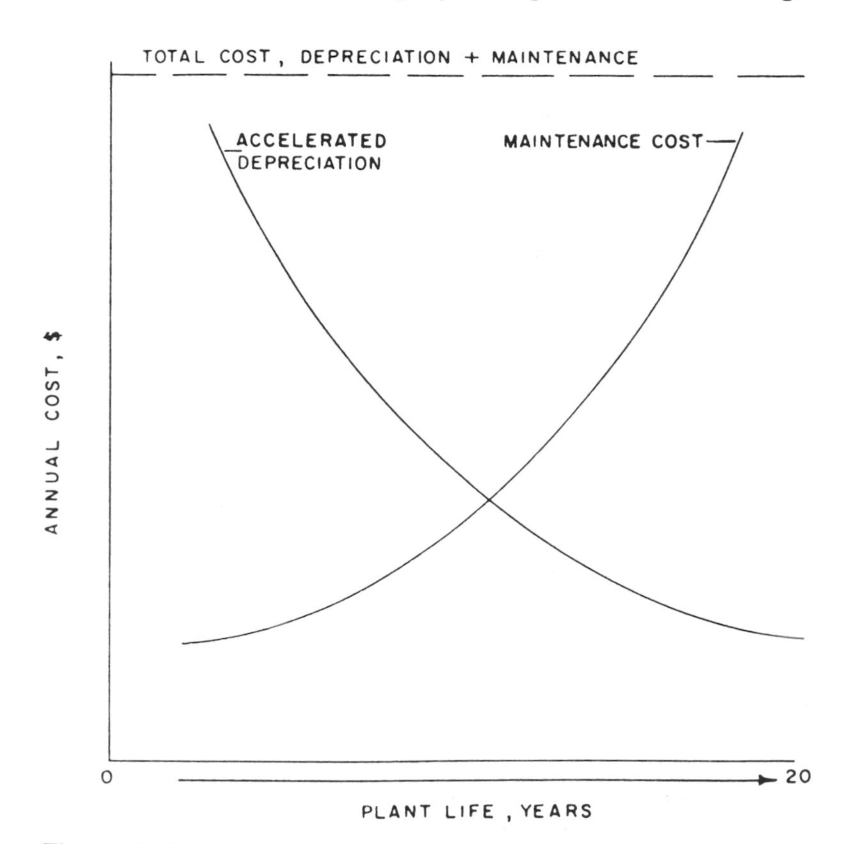

Figure 10.9 Comparison of maintenance and depreciation costs over the life of a plant.

OPERATING SUPPLIES AND OVERHEAD COSTS

Generally operating (or factory) supplies are a relatively minor cost of operations. Nevertheless these must be included in the operating or manufacturing cost estimate. Such costs include miscellaneous items such as lubricating oil, instrument charts, wiping cloths, etc. Lacking more detailed information, they may be estimated as a percentage of payroll. This percentage can range from a few percent to 20% or more, depending upon plant complexity and whether or not routine maintenance items and supplies are included or accounted for separately. For example, 6% of payroll is probably an adequate allowance for operating supplies in a coal preparation plant while 20% is probably more reasonable for an oil refinery, a more complex operation requiring a cleaner environment. The best source of such costs, however, is always company records of similar past projects.

Overhead or burden costs (see Appendix A) are operating and manufacturing costs which, while not directly proportional or related to production, are associated with payroll or general and administrative expense. Such costs, depending upon what they represent, are either semivariable or indirect costs.

The major semivariable overhead costs are so-called payroll overheads. These are costs which are associated with employee "fringe benefits." They include workers' compensation, pensions, group insurance, paid vacations and holidays, Social Security, unemployment taxes and benefits, profit-sharing programs, and a host of others. The extent of these costs varies markedly from industry to industry and company records are the best measure of their magnitude. However, in the absence of company data, payroll overheads may be roughly estimated at 25–40% of direct labor plus supervision plus maintenance labor costs for the United States. For other countries this factor must be adjusted to suit local conditions. In heavily socialized nations, payroll overheads can exceed 100% of the basic labor costs.

In addition, payroll overhead must be applied to indirect overhead (clerical, administrative, and other personnel) if not previously included in cost estimates for this item.

It must be noted that, if company data are used, care must be exerted to avoid including items in payroll overhead which are, by accounting definition of the company, included in general expense. Further, it should be recognized that in some industries, notably the U.S. and Canadian coal mining industry, a major portion of fringe benefits is based upon royalties levied by the unions on production rather than being a function of labor costs. Such royalties are variable production costs and must be treated as such.

The expense of operating company testing and research laboratories is another overhead expense which must be included in the estimate. Generally, such costs are indirect costs, although in the case of product laboratories they may be direct semivariable costs. Laboratory overhead is best estimated based upon company experience. Lacking suitable data, these costs can be estimated as follows:

From workhours required plus associated overhead
From literature sources
As a percentage of direct labor costs

If the last is used, laboratory overhead costs may range from 3–20% or more for complex processes. A suitable figure for average situations might be 5–10%.

ROYALTIES AND RENTALS

As was discussed earlier in this chapter, royalties may be variable, semivariable, fixed, or capital costs (or a combination of these) depending upon the conditions of the royalty agreement. The same is true of rental costs.

Single-sum royalty, rental, or license payments are properly considered as capital investment items, whereas payments in proportion to production or fixed payments per annum are treated as direct operating costs.

Royalty expenses, in the absence of data to the contrary, are treated as a direct expense and may be estimated at 1–5% of the product sales price.

Due to the complexity of agreements for royalty payments and to variations in tax laws and accounting methods, extreme care should be exercised to be sure that such costs are properly included in the appropriate expense category.

CONTINGENCIES

As is true with a capital cost estimate, any operating or manufacturing cost estimate should include a contingency allowance to account for those costs which cannot readily be determined or defined or which are too small to estimate individually but may be significant in the aggregate. The contingency allowance applies both to direct and indirect costs and ranges from 1–5% (and more in some cases) depending upon the uncertainty in the data used to prepare the estimate and the risk associated with the venture.

Hackney (5) has suggested the following guidelines for contingency allowances in operating and manufacturing cost estimates:

1. Installations similar to those currently used by the company, for which standard costs are available: 1%
2. Installations common to the industry, for which reliable data are available: 2%
3. Novel installations that have been completely developed and tested: 3%
4. Novel installations that are in the development stage: 5%

GENERAL WORKS EXPENSE

General works expense or factory overhead represents the indirect cost of operating a plant or factory and is dependent upon both investment and labor. Black (1) suggested that factory overhead be estimated by the sum of investment times an investment factor and labor times a labor factor. In this case, labor is defined as total annual cost of labor including direct operating labor, repair and maintenance, and supervision; and labor for loading, packaging, and shipping.

Black's suggested labor and investment factors for various industries are as follows:

Industry	Investment factor (%/year)	Labor factor (%/year)
Heavy chemical plants, large-capacity	1.5	45
Power plants	1.8	75
Electrochemical plants	2.5	45
Cement plants	3.0	50
Heavy chemical plants, small capacity	4.0	45

Alternately, for preliminary estimates, indirect overhead may be approximated at 40–60% of labor costs, or 15–30% of direct costs. Humphreys (6) has suggested 55% of operating labor, supervision, and maintenance labor for the mineral industries. Again, these factors may be somewhat higher outside the United States depending upon local customs and laws.

It is important to note that indirect or factory overhead (general works expense) *does not* include so-called general expense (i.e., marketing or sales cost) and administrative expense.

DEPRECIATION

Depreciation, while not a true operating cost, is considered to be an operating cost for tax purposes. It is customarily listed as a fixed, indirect cost. The purpose of depreciation is to allow a credit against operating costs, and hence taxes, for the nonrecoverable capital expense of an investment. The basis for computation of depreciation is the total initial capital expense for tangible assets, including interest during construction and start-up expense. The depreciable portion of capital expense is equal to the total initial investment less working capital and salvage value as shown in Figure 10.1.

In theory, working capital can be totally recovered at any time after the plant or process is shut down. Similarly, salvage value, the scrap or sales value of the process equipment at the end of its useful life, can in theory be recovered at any time after plant shutdown. Thus, the sunk and permanently lost capital is the total initial investment less working capital and salvage value (including land value). Through depreciation, this sunk investment may be recovered as an operating expense over the useful life of the project.

Unfortunately, true useful life of a project generally does not correlate with the permissible depreciation period dictated by tax laws. The United States Internal Revenue Service (IRS) establishes criteria for useful life of various investments which must be observed in cost and tax calculations whether or not these criteria actually reflect the true projected life of the plant or process. Table 10.3 lists typical permissible depreciation periods (the class life) for various plants and investments as approved by the Internal Revenue Service. In countries other than the United States, local taxing authorities should be consulted to determine the permissible life which can be used in depreciation calculations for any particular type of investment.

Taxing authorities usually permit the use of any generally accepted method of depreciation calculation *provided that* it is applied in a consistent manner to all investments applicable to the plant or process being considered. Different depreciation techniques may not be applied to various portions of the total investment. Further, effective in 1981 in the United States, a specialized system known as the *accelerated cost recovery system (ACRS)* was mandated by law. Subsequently, in 1986, the U.S. tax laws were revised again, and the ACRS system was replaced by a system called the *modified accelerated cost recovery system (MACRS)*. Both systems are described later in this chapter.

Most industrial firms utilize accelerated depreciation in their actual operating cost and tax calculations. Such techniques permit a major portion of the investment to be deducted from costs in the early years of the life of the plant, thus deferring taxes to the latest possible date. However, for the purpose of making preliminary operating and manufacturing cost estimates, straight-line depreciation is normally used *even though* the company may in fact use accelerated depreciation in its books. The reason for this apparent anomaly is that, as explained in the previous discussion, maintenance costs,

Table 10.3 Depreciation Class Lives and Recovery Periods

	Recovery periods (in years)		
	Class life	GDS (MACRS)	ADS
Specific depreciable assets used in all business activities, except as noted:			
Office furniture, fixtures, and equipment	10	7	10
Information systems	6	5	5
Data handling equipment, except computers	6	5	6
Airplanes (airframes and engines), except those used in commercial or contract carrying of passengers or freight, and all helicopters (airframes and engines)	6	5	6
Automobiles, taxis	3	5	5
Buses	9	5	9
Light general purpose trucks	4	5	5
Heavy general purpose trucks	6	5	6
Railroad cars and locomotives, except those owned by railroad transportation companies	15	7	15
Tractor units for use over-the-road	4	3	4
Trailers and trailer-mounted containers	6	5	6
Vessels, barges, tugs, and similar water transportation equipment, except those used in marine construction	18	10	18
Land improvements	20	15	22
Industrial steam and electric generation and/or distribution systems	22	15	22
Depreciable assets used in the following activities:			
Agriculture	10	7	10

Table 10.3 (*Continued*)

	Recovery periods (in years)		
	Class life	GDS (MACRS)	ADS
Cotton ginning assets	12	7	12
Cattle, breeding or dairy	7	5	7
Any breeding or work horse that is 12 years old or less at the time it is placed in service	10	7	10
Any breeding or work horse that is more than 12 years old at the time it is placed in service	10	3	10
Any race horse that is more than 2 years old at the time it is placed in service	None	3	12
Any horse that is more than 12 years old at the time it is placed in service and that is not a race horse, breeding horse, nor a workhorse	None	3	12
Any horse not described above	None	7	12
Hogs, breeding	3	3	3
Sheep and goats, breeding	5	5	5
Farm buildings except single-purpose agricultural or horticultural structures	25	20	25
Single-purpose agricultural or horticultural structures (GDS = 7 years before 1989)	15	10	15
Mining	10	7	10
Offshore drilling	7.5	5	7.5
Drilling of oil and gas wells	6	5	6
Exploration for and production of petroleum and natural gas deposits	14	7	14
Petroleum refining	16	10	16
Construction	6	5	6

Table 10.3 (*Continued*)

	Recovery periods (in years)		
	Class life	GDS (MACRS)	ADS
Manufacture of grain and grain mill products	17	10	17
Manufacture of sugar and sugar products	18	10	18
Manufacture of vegetable oils and vegetable oil products	18	10	18
Manufacture of other food and kindred products	12	7	12
Manufacture of food and beverages—special handling devices	4	3	4
Manufacture of tobacco and tobacco products	15	7	15
Manufacture of knitted goods	7.5	5	7.5
Manufacture of yarn, thread, and woven fabric	11	7	11
Manufacture of carpets and dyeing, finishing, and packaging of textile products and manufacture of medical and dental supplies	9	5	9
Manufacture of textured yarns	8	5	8
Manufacture of nonwoven fabrics	10	7	10
Manufacture of apparel and other finished products	9	5	9
Cutting of timber	6	5	6
Sawing of dimensional stock from logs, permanent or well-established	10	7	10
Sawing of dimensional stock from logs, temporary	6	5	6
Manufacture of wood products and furniture	10	7	10

Table 10.3 (*Continued*)

	Recovery periods (in years)		
	Class life	GDS (MACRS)	ADS
Manufacture of pulp and paper	13	7	13
Manufacture of converted paper, paperboard, and pulp products	10	7	10
Printing, publishing, and allied industries	11	7	11
Manufacture of chemicals and allied products	9.5	5	9.5
Manufacture of rubber products	14	7	14
Manufacture of rubber products—special tools and devices	4	3	4
Manufacture of finished plastic products	11	7	11
Manufacture of finished plastic products—special tools	3.5	5	3.5
Manufacture of leather and leather products	11	7	11
Manufacture of glass products	14	7	14
Manufacture of glass products—special tools	2.5	3	2.5
Manufacture of cement	20	15	20
Manufacture of other stone and clay products	15	7	15
Manufacture of primary nonferrous metals	14	7	14
Manufacture of primary nonferrous metals—special tools	6.5	5	6.5
Manufacture of foundry products	14	7	14
Manufacture of primary steel mill products	15	7	15
Manufacture of fabricated metal products	12	7	12
Manufacture of fabricated metal products—special tools	3	3	3

Table 10.3 (*Continued*)

	Recovery periods (in years)		
	Class life	GDS (MACRS)	ADS
Manufacture of electrical and non-electrical machinery and other mechanical products	10	7	10
Manufacture of electronic components, products, and systems	6	5	6
Any semiconductor manufacturing equipment	5	5	5
Manufacture of motor vehicles	12	7	12
Manufacture of motor vehicles—special tools	3	3	3
Manufacture of aerospace products	10	7	10
Ship and boat building machinery and equipment	12	7	12
Ship and boat building dry docks and land improvements	16	10	16
Ship and boat building—special tools	6.5	5	6.5
Manufacture of locomotives	11.5	7	11.5
Manufacture of railroad cars	12	7	12
Manufacture of athletic, jewelry and other goods	12	7	12
Railroad transportation:			
Railroad machinery and equipment	14	7	14
Railroad structures and similar improvements	30	20	30
Railroad wharves and docks	20	15	20
Railroad track	10	7	10
Railroad hydraulic electric generating equipment	50	20	50

Table 10.3 (*Continued*)

	Recovery periods (in years)		
	Class life	GDS (MACRS)	ADS
Railroad nuclear electric generating equipment	20	15	20
Railroad steam electric generating equipment	28	20	28
Railroad steam, compressed air, and other power plant equipment	28	20	28
Motor transport—passengers	8	5	8
Motor transport—freight	8	5	8
Water transportation	20	15	20
Air transport	12	7	12
Air transport (restricted)	6	5	6
Pipeline transportation	22	15	22
Telephone communications			
Telephone central office buildings	45	50	45
Telephone central office equipment	18	10	18
Computer-based telephone central office switching equipment	9.5	5	9.5
Telephone station equipment	10	7	10
Telephone distribution plant	24	15	24
Radio and television broadcastings	6	5	6
Telegraph, ocean cable, and satellite communications (TOCSC)			
TOCSC—Electric power generating and distribution systems	19	10	19
TOCSC—High frequency radio and microwave systems	13	7	13
TOCSC—Cable and long-line systems	26.5	20	26.5
TOCSC—Central office control equipment	16.5	10	16.5

Table 10.3 (*Continued*)

	Recovery periods (in years)		
	Class life	GDS (MACRS)	ADS
TOCSC—Computerized switching, channeling, and associated control equipment	10.5	7	10.5
TOCSC—Satellite ground segment property	10	7	10
TOCSC—Satellite space segment property	8	5	8
TOCSC—Equipment installed on customer's premises	10	7	10
TOCSC—Support and service equipment	13.5	7	13.5
Cable television (CATV)			
CATV—Headend	11	7	11
CATV—Subscriber connection and distribution systems	10	7	10
CATV—Program origination	9	5	9
CATV—Service and test	8.5	5	8.5
CATV—Microwave systems	9.5	5	9.5
Electric, gas, water and steam, utility services			
Electric utility hydraulic production plant	50	20	50
Electric utility nuclear production plant	20	15	20
Electric utility nuclear fuel assemblies	5	5	5
Electric utility steam production plant	28	20	28
Electric utility transmission and distribution plant	30	20	30
Electric utility combustion turbine production plant	20	15	20
Gas utility distribution facilities	35	20	35
Gas utility manufactured gas production plants	30	20	30

Table 10.3 (*Continued*)

	Recovery periods (in years)		
	Class life	GDS (MACRS)	ADS
Gas utility substitute natural gas (SNG) production plant (naphtha or lighter hydrocarbon feedstocks)	14	7	14
Substitute natural gas-coal gasification	18	10	18
Natural gas production plant	14	7	14
Gas utility trunk pipelines and related storage facilities	22	15	22
Liquefied natural gas plant	22	15	22
Water utilities	50	20	50
Central steam utility production and distribution	28	20	28
Waste reduction and resource recovery plants	10	7	10
Municipal wastewater treatment plant	24	14	24
Municipal sewer	50	20	50
Distributive trades and services	9	5	9
Distributive trades and services—billboard, service station buildings and petroleum marketing land improvements	20	15	20
Recreation	10	7	10
Theme and amusement parks	12.5	7	12.5

Notes: In those cases where guidelines are not listed for any given industry or type of equipment, or where the listed guidelines are clearly inappropriate, the depreciable life of such property shall be determined according to the particular facts and circumstances.

Source: "Depreciation," U.S. Department of the Treasury, Internal Revenue Service, Publication No. 534, 1994.

which are known to increase with time, are generally assumed to be constant with time. As shown in Figure 10.9, accelerated depreciation increases with time. Assuming both to be constant, i.e., assuming straight-line depreciation, generally results in offsetting errors, as the actual sum of the two factors, for most plants, tends to be constant, or essentially constant, over the life of any given plant.

To calculate straight-line depreciation, annual depreciation is simply made equal to the depreciable portion of the initial capital investment divided by the depreciable life of the project.

In the case of projects with components having different depreciable lives, each component may be depreciated separately or the weighted average life may be used on the total depreciable investment.

Mathematically, annual straight-line depreciation is equal to

$$D_{sl} = \frac{C}{Y}$$

where

D_{sl} = annual straight-line depreciation

C = depreciable portion of capital investment

Y = IRS-approved life, in years

There are many other acceptable depreciation techniques, of which the double-declining balance and sum-of-years-digits methods, both forms of accelerated depreciation, are the most commonly used. In the double-declining balance method, an annual depreciation deduction is permitted on the undepreciated portion of the investment at a rate equal to twice the straight-line rate. For example, if an investment has an approved life of 5 years for depreciation purposes, the straight-line deduction is 20% of the original depreciable investment per year. Thus, the double-declining balance deduction is twice this rate, or 40%.

An example with an investment of $1 million, a salvage value of zero, and a 5–yr life, annual double-declining balance depreciation allowances follows:

Year	Investment at 40%	Depreciation ($)	Undepreciated balance
0	$1,000,000	—	$1,000,000
1	1,000,000	$400,000	600,000
2	600,000	240,000	360,000
3	360,000	144,000	216,000
4	216,000	86,400	129,600
5	129,600	129,600[a]	0
Total		$1,000,000	

[a] In the final year of the life, the total remaining undepreciated balance may be deducted.

Mathematically, the double-declining balance method is

$$D = \frac{2(F - CD)}{n}$$

where

D = depreciation in any given year

F = initial asset value

CD = cumulative depreciation charged in prior years

n = asset life

Note that in the double-declining balance method, the basis for depreciation ordinarily is the total depreciable investment *without* deducting for salvage values other than land value. In this method, the investment is depreciated down to the salvage value, and the final undepreciated balance may not be less than salvage value.

Another common method of computing accelerated depreciation is the "sum-of-years-digits" method. This technique is based upon the depreciable portion of the investment, i.e., excluding land *and* salvage value.

To use this technique, the approved years in the life of the plant are summed. Deductions are based on the remaining years of plant life divided by the sum of years. For example, with a 5–yr life, the sum of years digits equals 5 + 4 + 3 + 2 + 1 = 15. In the first year 5/15 of the depreciable investment is deducted; 4/15 in the second year, 3/15 in the third year; etc.

Using the same example as given above for the double-declining balance method, sum-of-years-digit depreciation deductions would be as follows:

Year	Investment	Depreciation ($)	Undepreciated balance
0	$1,000,000	—	$1,000,000
1	1,000,000	5/15 = 333,333	666,667
2	1,000,000	4/15 = 266,667	400,000
3	1,000,000	3/15 = 200,000	200,000
4	1,000,000	2/15 = 133,333	66,667
5	1,000,000	1/15 = 66,667	0
Total		$1,000,000	

The sum-of-years-digits method is expressed mathematically as

$$D_y = C\left[\frac{2(n-Y+1)}{n(n+1)}\right]$$

where

D_y = depreciation in year Y

C = depreciable portion of investment

n = asset life, in years

There are numerous other acceptable depreciation methods including some which are combinations of the above methods. However, the three methods described above are the most commonly used techniques.

Also as mentioned earlier, if constant maintenance costs are assumed, no matter which depreciation technique is actually used by the company, the straight-line technique should be generally used for are all preliminary estimates.

ACCELERATED COST RECOVERY SYSTEM

In 1981 a major revision of tax laws in the United States replaced the pre-existing depreciation systems described above with a system known as the accelerated cost recovery system (ACRS). ACRS is mandatory for all capital assets acquired after 1980 and before 1987 when another system of depreciation, the modified accelerated cost recovery system (MACRS) became mandatory. However, the tax laws specify that any acquisition must continue to be depreciated on its original basis. Thus, since many capital assets have depreciable lives of up to 60 years, the old depreciation systems, MACRS, and ACRS will coexist and be used by cost professionals for many years into the future.

Under ACRS, capital assets are not subject to depreciation in the customary sense. It is not necessary to estimate salvage values or useful lives for equipment. Instead, the law establishes various property classes and provides for deductions calculated as specific percentages of the cost of the asset. Property classes are:

1. *Three-year property*, which is defined as "property that has a midpoint class life of four years or less, or, is used for research and experimentation, or, is a race horse more than two years old when placed in service, or any other horse that is more than 12 years old when placed in service." This obscure and somewhat confusing-sounding definition includes automobiles, light trucks, and shortlived personal property.
2. *Five-year property*, which is defined as property not otherwise defined and which is not real property. This class includes most types of equipment and machinery.
3. *Ten-year property*, which is public utility property having a midpoint class life of more than 18 but not more than 25 years. Also included are manufactured homes, railroad tank cars, certain coal utilization property, theme and amusement park property, and other property as defined in the act.
4. *Fifteen-year property*, which is long-lived public utility property.
5. *Ten-year and fifteen-year real property classes*. The 15–year class consists of real property with a midpoint class life in excess of 12.5 years. All other real property falls into the 10–year class.

Detailed descriptions of each property class may be obtained upon request from the U.S. Internal Revenue Service.

Under the 1981 law, ACRS deductions were phased in on a gradual basis over a period of years. Allowable deductions for personal property are listed in Table 10.4. Fifteen-year class real property deductions vary from Table 10.4 and are based in part on the month in which the property was

Table 10.4 ACRS Deductions (%) for Personal Property and Ten-Year Real Property

	Year property was put in service		
	1981–1984	1985	1986
3–year property			
1st year	25	29	33
2nd year	38	47	45
3rd year	37	24	22
5–year property			
1st year	15	18	20
2nd year	22	33	32
3rd year	21	25	24
4th year	21	16	16
5th year	21	8	8
10–year property			
1st year	8	9	10
2nd year	14	19	18
3rd year	12	16	16
4th year	10	14	14
5th year	10	12	12
6th year	10	10	10
7th year	9	8	8
8th year	9	6	6
9th year	9	4	4
10th year	9	2	2

Table 10.4 (*Continued*)

	Year property was put in service		
	1981–1984	1985	1986
15–year personal property			
1st year	5	6	7
2nd year	10	12	12
3rd year	9	12	12
4th year	8	11	11
5th year	7	10	10
6th year	7	9	9
7th year	6	8	8
8th year	6	7	7
9th year	6	6	6
10th year	6	5	5
11th year	6	4	4
12th year	6	4	3
13th year	6	3	3
14th year	6	2	2
15th year	6	1	1

placed in service. Internal Revenue Service regulations should be consulted to determine applicable deduction rates.

It should also be noted that the ACRS requirements place certain limitations on deductions for disposition of property prior to the end of its class life, and also provide for optional use of alternate percentages based on the straight-line method of depreciation.

The logic of the standard ACRS percentages as outlined in Table 10.4 becomes apparent when it is realized that the 1981 to 1984 rates are approximately equal to those for 150% declining-balance depreciation with a switch to straight-line depreciation in later years. The 1985 rates similarly approximate 175% declining-balance depreciation, and the rates for 1986 are essentially those for the double-declining-balance method, both with a switch to the sum-of-years-digits method at the optimum point in time. Thus ACRS is merely a combination of previously used and widely accepted methods of computing accelerated depreciation.

MODIFIED ACCELERATED COST RECOVERY SYSTEM

When the ACRS depreciation system was adopted in the United States, it was phased in over a period of five years ending in 1986 and was originally anticipated to remain in effect after that time. The 1986 tax reform act, however, further revised the depreciation regulations and, effective in 1987, implemented a new system called the *modified accelerated cost recovery system* (MACRS).

MACRS expanded the ACRS property classes from 5 to 8, revised the depreciation periods for most items (see Table 10.3), and redefined the classes as follows:

1. *Three-year property*, which was redefined as including "tractor units for use over the road, any race horse over 2 years old when placed in service, and any other horse over 12 years old when placed in service." Three-year property generally includes those items with an IRS-approved class life of 4 years or less. The major change in this class was the elimination of automobiles, light trucks, and other shortlived personal property which was moved to the 5–year class.
2. *Five-year property*, which was totally redefined as including "trucks, computers and peripheral equipment, office machinery, and any automobile." Most items with an IRS-approved class life of 4–plus years and less than 10 years are included under this category.

3. *Seven-year property*, which is defined in part as including "office furniture and fixtures, any property that does not have a class life, and that has not been designated by law as being in any other class and, if placed in service before 1989, any single purpose agricultural or horticultural structure." This category covers any items with an IRS-approved class life of 10 years or more and less than 16 years. It effectively includes almost all industrial machinery and equipment.
4. *Ten-year property*, which includes "vessels, barges, tugs and similar water transportation equipment, and, if placed in service after 1988, any single-purpose agricultural or horticultural structure, and any tree or vine bearing fruit or nuts. This category includes those items with an IRS-approved class life of 16 years or more and less than 20 years.
5. *Fifteen-year property*, which includes items with an IRS-approved class life of 20 years or more and less than 25 years plus wastewater treatment plants and equipment used for two-way exchange of voice and data communications.
6. *Twenty-year property*, which includes items with an IRS-approved class life of 25 years or more, excluding real property and including sewer systems. Farm buildings (other than single-purpose agricultural or horticultural structures) fall into this category.
7. *Twenty-seven and one-half year property*, which includes residential rental property.
8. *Nonresidential real property*, which includes real property other than residential rental property. For this category, the recovery period for depreciation is 31.5 years for property placed in service *before* May 13, 1993 and 39 years for property placed in service after May 12, 1993.

The depreciation allowances under the MACRS system, like ACRS, do not consider salvage value. For the 3-, 5-, 7-, and 10-year categories, the MACRS depreciation schedule (Table 10.5) is equivalent to 200% declining-balance switching to straight-line at the optimum point to maximize the deduction. In the first year, a half-year convention applies. The assumption is made that the item being depreciated was placed in service for only 6 months no matter what the actual date of service was. However, if more than 40% of the cost basis was placed in service during the last 3 months of the year, a mid-quarter convention applies, i.e., it is assumed that the equipment was in service for only 1.5 months of the first year.

Table 10.5 MACRS Deduction Rates

If the recovery year is:	3–year	5–year	7–year	10–year	15–year	20–year
1	33.33	20.00	14.29	10.00	5.00	3.750
2	44.45	32.00	24.49	18.00	9.50	7.219
3	14.81	19.20	17.49	14.40	8.55	6.677
4	7.41	11.52	12.49	11.52	7.70	6.177
5		11.52	8.93	9.22	6.93	5.713
6		5.76	8.92	7.37	6.23	5.285
7			8.93	6.55	5.90	4.888
8			4.46	6.55	5.90	4.522
9				6.56	5.91	4.462
10				6.55	5.90	4.461
11				3.28	5.91	4.462
12					5.90	4.461
13					5.91	4.462
14					5.90	4.461
15					5.91	4.462
16					2.95	4.461
17						4.462
18						4.461
19						4.462
20						4.461
21						2.231

For the 15- and 20-year categories, the depreciation schedule, as shown in Table 10.5, is 150% declining-balance with a switch to straight-line at the optimum point. The half-year and half-quarter conventions also apply to these categories.

For the real property categories, straight-line depreciation must be used with a mid-month convention in the first year.

If all of the foregoing seems confusing and convoluted, it is. The logic of tax laws and their complexity is rarely clear, and the MACRS system is a significant example of how the political process can complicate what should otherwise be an easily understood subject.

To further complicate the calculation of depreciation, the MACRS system allows for an alternate method. As shown in Table 10.3, the alternate depreciation system (ADS) permits straight-line depreciation over specified periods which are equal to, or longer than, the regular MACRS recovery periods (GDS, the general depreciation system). Generally, the taxpayer must specifically choose to use the alternate method; otherwise the regular MACRS system applies. In a very few cases, the alternate system must be used. The exceptions are small in number and generally are not of concern to the cost engineer.

The alternate system, however, is always a bad choice economically. The regular system permits far more rapid deductions and resultant tax savings. The alternate system is easier to understand and is simple to calculate but is a poor choice—it costs money. Nevertheless, the government permits poor economic decisions. The wise cost engineer should avoid them.

AMORTIZATION, DEPLETION, INSURANCE, AND REAL ESTATE TAXES

Amortization is a term which is applied to writing off or recovering any portion of the initial capital expense which is intangible in nature and as such has no definable useful life. A lump-sum royalty payment is an example of such an investment. Such intangible assets are written off as an operating cost over the life of the plant or process using exactly the same calculation techniques as are used for depreciation of tangible assets. Alternately, intangible assets may be written off as a function of production over the life of a project. In some cases amortization is, for simplicity, included in the depreciation charge (albeit erroneously). However, for practical purposes the distinction between depreciation and amortization is usually of no consequence.

Depletion allowances, while not considered to be an operating cost, must be included in estimates involving extraction of a natural resource, e.g., in coal mining. These allowances are deductions from gross income prior to calculation of taxes on income. Thus they are, in effect, tax credits granted by law to compensate for eventual exhaustion of an irreplaceable natural resource such as coal or oil. They are computed as a fixed percentage of the market value of the resource in its *first usable and salable form*, even though the resource may be further processed and eventually sold in another form at a different cost. Depletion rates are established by law and are periodically changed. Thus it is necessary when performing an estimate first to determine the currently applicable rate.

Insurance and real estate (or property) taxes must also be included in the estimate if not previously considered in determining general works expense. In most areas these costs total about 1.5–3% of investment per year. As stated in Chapter 9, 2% is about average for locations in the United States.

DISTRIBUTION COSTS

The costs of packing and shipping products to market (i.e., distribution costs), are highly variable and are dependent upon product characteristics. In many cases, especially with consumer products, these costs often exceed the cost of producing the product itself. Distribution costs may include the following:

1. Cost of containers, including their repair, testing, cleaning, etc., if reusable, and their depreciation or rental, if nonexpendable
2. Transportation costs
3. Applicable labor and overheads for packing and shipping

If the product is sold FOB the plant, the cost of transportation is borne by the customer and need not be considered in the estimate. If, however, it is sold on a delivered basis, transportation costs must be included.

The mode of transportation and the shipping distance drastically affect transportation costs. In general, pipelines, barges, and tankers are the least expensive forms of transportation. Rail shipment is somewhat higher in cost, and truck shipment is the most expensive. Freight tariffs are regulated by the states (intrastate shipments) and the federal government (interstate shipments) and vary from product to product. In general, for any given product, freight costs per ton-mile are lowest for long hauls (250–300 miles or more) and high volumes. Short-distance haulage and small shipments incur considerably higher freight rates.

PROBLEMS AND QUESTIONS

10.1 a) Define fixed cost, variable cost, and semivariable cost. What items are included in each? b) At zero production, what percentage of semivariable cost is normally incurred (as a percentage of total semivariable cost at full production)? c) What are the definitions of the breakeven and shutdown points?

10.2 In a manufacturing operation, at 100% of capacity, annual costs are as follows:

Fixed expense	$4,730,400
Variable expense	6,446,400
Semivariable expense	5,652,500
Sales	23,986,800

Assuming that semivariable costs at zero production equal 30% of such costs at 100% of capacity, determine the shutdown and breakeven points.

10.3 You work for a chemical company which plans to install $60,000,000 worth of new equipment this year. You estimate that this equipment will have a salvage value of $5,000,000 at the end of its useful life. What are the Internal Revenue Service (IRS) approved depreciation allowances for the first five years?

10.4 In the above problem, what is the *straight line* depreciation for each of the first five years? This technique is recommended for preliminary estimates. Why?

10.5 Assume that for accounting purposes rather than tax purposes, your company uses double declining balance depreciation. Calculate the double declining balance depreciation allowances for each of the first five years of equipment life for problem 10.3.

10.6 What is a depletion allowance? How is it calculated?

10.7 What are general and administrative expenses?

10.8 What is general works expense?

10.9 Why should operating cost estimates be made on an annual basis rather than a daily or unit-of-production basis?

10.10 How can the cost of grease and lubricants be estimated for motor vehicles and mobile equipment?

10.11 How can fuel cost be estimated for motor vehicles and mobile equipment?

10.12 In order to estimate process labor requirements, what must first be established?

10.13 How many standard 40–hour/week shifts are required to staff a plant around the clock, 7 days a week?

10.14 What is a good rule of thumb for estimating electric power costs?

10.15 What rule of thumb can be used to estimate the cost of steam?

10.16 In scaling up utililty requirements for larger pieces of equipment, what rule of thumb can be used to estimate the utility requirements?

10.17 Is it advisable to use scheduled overtime work in industrial plants?

10.18 If a plant operates at 50% of capacity, what will its approximate maintenance cost be as a percentage of maintenance costs at full capacity.

10.19 For preliminary estimates, annual maintenance costs are generally assumed to be constant even though they are known to increase with time. Straight-line depreciation is also assumed despite the fact that accelerated depreciation will be used in actual operations. Why are these two clearly erroneous assumptions made?

10.20 Are royalties and rental costs variable, semivariable, fixed, or capital costs?

REFERENCES

1. J.H. Black, Operating cost estimation, in *Jelen's Cost and Optimization Engineering* (K.K. Humphreys, ed.), McGraw-Hill, New York, 1991, Chap. 15.
2. H.E. Wessell, New graph correlates operating labor data for chemical processes, *Chemical Engineering*, 209 (July 1952).
3. R.M. Blough, Effect of scheduled overtime on construction projects, *AACE Bulletin 15*(5), 155–158, 160 (October 1973).
4. W. McGlaun, Overtime in construction, *AACE Bulletin 15*(5), 141–143 (October 1973).
5. J.W. Hackney, Estimate production costs quickly, *Chemical Engineering*, 183 (April 17, 1971).
6. K.K. Humphreys, Coal preparation costs, in *Coal Preparation* (J.W. Leonard, ed.), 5th ed., Society of Metallurgy and Exploration, Littleton, CO, 1992, Chap. 3.

11

Transition
Construction to Operation

There is, in the normal evolution of a project, a relatively long transition period where costs are incurred, and it is not always clear whether these costs should be charged to construction or operations. These costs are not addressed in most textbooks on estimating but can be a significant expenditure.

On almost all major projects there is an owner's management team which is not included in the construction contract; the owner normally wants key operating personnel brought into a project early. The project management team may be a small group of highly specialized people for a routine-type project to a large specialized team for a megaproject varying from 5–10 up to 150 personnel. The project team is in residence at the engineering and construction firm's offices at the beginning of the project. The team then goes to the field for actual plant construction. It is imperative for the project team, owner, and contractor, to work together to achieve the common goal of bringing the project in under budget and on time. It is easy to let relations degrade to a confrontational situation which is undesirable for both the owner and the contractor. Thus, the personnel chosen for the project management

Table 11.1 Project ABC Pre-operating, Training, Startup, and Debugging Personnel (by year)

Project year	Hourly personnel	Salaried personnel	Startup and debugging	Construction project team
1	5	21		45
2	38	42		68
3	52	90		80
4	87	122		89
5	212	148		94
6	612	151		94
7	925[a]	160[a]	10	69
8			10	25
9			25	

[a]Covered in operating-cost table, first-year operations.

team on a construction job must be competent, tactful people to assure harmonious execution of the contract. One obnoxious person in such a situation can be catastrophic to the timely completion of the work.

For large corporations, the owner's cost is normally expensed as spent; however, it might, under unusual circumstances, be advantageous from tax considerations to capitalize the costs. The decision as to how the costs are treated should be made in consultation with a tax expert at the time the decision is needed.

Table 11.1 is a summary of owner's management personnel and costs taken from an actual megaproject. The first two columns of Table 11.1 are the operations people brought into the project early and include the key personnel for operations. The third column comprises the debugging people retained from the plant construction to aid in plant startup. The last column is the owner's construction project team, of which many members could be retained for operations if desired.

The operations people perform a number of important functions other than becoming familiar with the plant processing and construction. They have the responsibility of writing the operating manuals and procedures for startup,

Table 11.2 Project ABC Preoperating and Startup Expenses

Project year	Total 1995 dollars
1	7,203,000
2	13,990,000
3	21,171,000
4	27,779,000
5	39,554,000
6	62,461,000
7	8,222,000
8	4,163,000
9	2,602,000
Total	$187,145,000

operation, and shutdown of all the various units in the processing complex. Prior to startup the operators must be trained, and this function is carried out by the operations people. Please note that for this project owner's construction cost and operations costs overlap in years 7 through 9. Plant startup is initiated in year 7.

The startup and debugging people are held over from construction (design people) to aid in debugging the equipment as each processing unit is brought onstream.

The owner's construction team peaks about two-thirds through the construction period and then diminishes after startup and disappears in the third year.

Table 11.2 shows the owner's estimated cost for each year and the total owner's costs for construction through startup. As may be seen, the expenses incurred for a megaproject are significant. Costs for a smaller project would be less, but these costs should not be omitted from the project evaluation.

Part III

Financial Analysis and Reporting

12

Profitability Analysis*
An Overview

CASH FLOW

While theoretically a project or process is intended to produce a product or service, the true goal of any industrial enterprise is to produce a profit. Funds must be expended wisely, resulting in a new gain to the company or the business will not survive.

The first step in determining the profitability of a process is to determine the cash flow into or out of the process for each year over its life. *Cash flow* is defined as "the net flow of dollars into or out of the proposed project—the algebraic sum, in any time period, of all cash receipts, expenses, and investments" (see Appendix A).

Upon completion of the operating cost estimate for a project or process, cash flow can readily be determined. The cash flow, for any given year is equal to:

---

* Adapted from articles authored by Julian A. Piekarski, PAICE Associates, Inc., Milwaukee, WI.

Net after tax income
plus Depreciation
plus Depletion
plus Any other tax credits claimed as an operating cost or as a deduction from gross income
less Any additional capital investments made during the year
plus Net income from sale or salvaging or used equipment or other assets
plus In the final year of the project's life, working capital

Net after tax income is equal to:

Net before tax income
less Federal, state, and local taxes on income (usually about 50%)

Net before tax income is equal to:

Gross income from sales
less Total operating cost, including depreciation
less Depletion claimed (if any)
less Any other applicable tax credits claimed as deductions from income

As an example, consider a process with annual sales income of $1,000,000 per year, operating costs of $800,000, depreciation of $200,000, and depletion of $50,000. In this example, cash flow is calculated as follows:

Sales income	$1,000,000
less Operating cost	–800,000
Subtotal (gross income):	200,00
less Depletion	–50,000
Subtotal (net before tax income):	150,000
less Taxes at 51%	–76,500
Subtotal (net after tax income):	73,500
plus Depreciation	200,000
plus Depletion	50,000
Total (cash flow):	323,500

In this example, the true net cash flow of the project for the year in question is $323,500 or more than 4 times the calculated after-tax income. This is true because depreciation represents partial recovery of an earlier investment, not an actual out-of-pocket expense for the year in question. Depletion, similarly, is a tax credit granted by law and also does not represent an out-of-pocket expense. Thus, it should be apparent that situations can and do exist where processes have a net loss for tax purposes but can in fact have a net positive cash flow.

The key to the methodology for conducting financial studies and demonstrating how they can be applied is the cash flow table which is the principal schedule to integrate the inputs of revenue, operating cost, and capital investments. The focus naturally is on the bottom line or cash flow since this is the place where the discounting is done to produce the financial results desired in terms of net present value (venture worth) or interest rate of return (discounted cash flow).

Basic engineering economic techniques that are applicable to a wide range of problems and are consistent with good accounting and financial practice are presented in this chapter. That a significant number of companies use these techniques is evidence of their applicability.

BASIC CONCEPTS OF PROFITABILITY ANALYSIS

The primary objective of financial analysis is to provide the basis for decision making. The preceding chapters have set the stage for this analysis by translating the problem into terms of project scope and process design. These in turn have provided the framework for developing the inputs of revenue, operating costs, and capital investment. The discussion of the techniques by which this data can be transformed into a workable format from which the financial analysis can now proceed. First a few basic concepts need to be discussed.

How Do We Solve Alternative Choice Problems?

Generally, the major task is to structure the problem using the scientific method as a logic base. The central question is that, given a proposed investment or expenditure, we must determine the base economic solution in terms of present value. The basic methodology is as follows:

1. Identify the significant alternatives.
2. Quantify all of the relevant money flows in and out of the project over time.
3. Construct a cash flow table for each alternative.
4. Compute the present worth of each alternative at the appropriate interest rate: "discount the cash flow."
5. Compute the discounted cash flow rate of return (DCF).

There are additional concepts that apply to project evaluation:

1. The alternatives considered must be mutually exclusive: the decision to implement one precludes the need to do the remaining alternatives.
2. Judgment still plays an important role and should give weight to factors that cannot be reduced to money terms.
3. Economic studies start from the moment of decision. Only present and future money flows are relevant. Disbursements that have already occurred prior to the time of decision are sunk costs, unless they have an impact on future flows.

What Type of Projects Are Encountered in Our Day-to-Day Operations?

The decisionmaker is confronted with two basic types of situations that require expenditure of funds:

1. *Mandatory expenditures*: These generally include machine breakdowns or emergencies, as well as necessary maintenance required to keep production lines operating. The decision to spend the money has already been made by necessity. All that remains is to determine the most economic choice from among the potential alternatives.
2. *Expenditures open to choice*: Here the decision to spend the money will be based solely on whether it can be justified by savings and/or revenues that result from the action. There are two classes of projects as follows:
   a. *Cost reduction.* This type of expenditure is justified by savings that come from: process improvements, reduced labor requirements, and productivity improvements such as increased output or reduced scrap that result from special equipment tools or machine replacement.
   b. *Revenue generating*. This centers around expenditures that bring in new revenues to the company resulting from: product improvement, plant expansion, and new product introduction.

Revenue-generating projects are the main topic of this discussion.

How Is Project Evaluation Related to Capital Budgeting?

The purpose of project evaluation is to identify the most economic choice from among mutually exclusive alternatives. This is done for each major project that is under consideration. However, all companies are confronted with a limitation as to the amount of capital that is available for any period of time and thus have the further problem of budgeting or allocating the available capital to the various possible uses or projects. This requires good judgment and must take into consideration the importance of each expenditure with respect to its impact on company operations. Most projects that fall into the classification of mandatory expenditures have a natural priority and are given a heavier weight in this selection process. If funds are left over beyond this point, the remaining projects can be ranked on the basis of their potential earning rate contribution or life cycle cost as a means of rationing the capital in this category. Chapter 12 covers this type in some detail.

What Kind of Money Do We Spend?

There are basically two kinds of money that we spend:

1. Money that qualifies as an operating expense in the current tax or accounting period.
2. Capital dollars that fall below the tax line. These dollars in turn fall into two categories:
   a. Capital dollars that are subject to depreciation, such as equipment and buildings.
   b. Capital dollars that are not depreciated, such as working capital and land.

ASSUMPTIONS AND DEFINITIONS

Before any calculations can proceed, several assumptions and definitions have to be given. Some of these assumptions have already been discussed in prior chapters, particularly those dealing with capital investment and operating costs.

1. *Time dimensions*
   a. Project life should correspond to the practical working life of the process being studied.
   b. Depreciation life is mandated by IRS guidelines and may not necessarily be the same as project life. If the project life is shorter than the depreciation life, appropriate measures have to be taken to account for the remaining book value by "writing it off" as part of the liquidation procedure.

c. Base year should represent the logical point from which to identify costs where current dollars are used. In many instances the base year coincides with the start of the operations.

2. *Base case.* The base case is a set of estimates, parameter values, and the outcomes of the analysis which are to be used as a reference datum.
3. *Depreciation.* From an accounting standpoint depreciation is the gradual conversion of the cost of an asset into expense. Specifically, it is the allocation of the cost of a particular plant or equipment asset to the periods in which services are received from that asset. Depreciation does not mean the physical deterioration of an asset nor does it mean the decrease in market value. From a tax standpoint, depreciation is an arbitrary schedule of capital recovery governed by guidelines established by the U.S. Internal Revenue Service or corresponding agency in other countries.
4. *Cost of capital.* The cost of capital is defined in American National Standards Z-94.0* as an interest rate percentage which expresses "... the overall estimated cost of investment capital at a given point in time, including both equity and borrowed funds." From the standpoint of a company, the cost of capital represents the minimum goal of attainment. In project evaluation the cost of capital has two main uses:
   a. As the discount rate for venture worth calculations;
   b. As the hurdle rate (minimum acceptable rate of return) against which the project's discounted cash flow rate of return is measured.

   For government agencies it is quite common to use the cost of borrowing money as the cost of capital. This, of course, assumes that the government body is able to raise enough money to finance all projects at the same rate.

   For any one company the cost of capital should represent the weighted cost of the principal sources of capital: debt and equity. This can be easily shown by the following tables. For an electric utility with a capital structure comprised of 50% new debt borrowed at 9% and 50% equity at 14%, the weighted cost of capital is computed as follows:

---

* *ANSI Standard Z94.0-1989, Industrial Engineering Terminology, Revised Edition,* Industrial Engineering and Management Press, Norcross, GA and McGraw-Hill, Inc., New York, 1989.

	Debt	Equity	Total
Capital structure	50%	50%	100%
Cost	9%	14%	
Weighted cost:			
Before tax	4.5	7.0	11.5
After 50% tax	2.3	7.0	9.3

For a typical non-utility business the only assumptions that change are for the composition of the capital structure: 30% debt and 70% equity. The computations for the weighted cost of capital are as follows:

	Debt	Equity	Total
Capital structure	30%	70%	100%
Cost	9%	14%	
Weighted cost:			
Before tax	2.7	9.8	12.5
After tax	1.4	9.8	11.2

From the above calculations it can be seen that the leveraging effect of debt gives the utility a lower cost of capital than the typical non-utility business.

5. *Impact of inflation.* Cost estimates for future projects that extend many years into the future are technically impossible to make with any degree of accuracy due to the unpredictable nature of cost escalation in the future. Estimates should be made in current dollars. Any assumptions as to future cost levels are of questionable reliability.

METHODS OF MEASURING PROFITABILITY

There are a great many methods utilized by industry to assess the profitability, or potential profitability, of a process. The most widely used techniques are:

Payback or payout time
Payback or payout time with interest
Return on investment

Return on average investment
Discounted cash flow (also known as interest rate of return)
Venture worth analysis (also known as incremental present worth)

Each technique has its own advantages and shortcomings, and often projects are ranked in different orders of profitability depending upon which technique is used. For this reason, it is wise when comparing two or more alternatives to examine their profitability by at least two different profitability measures.

Payback or payout time is defined as the time required for all cash flows to equal the original investment, i.e., the time necessary to recover the original investment. As an example, assume a plant with an initial investment of $1,000,000 (i.e., an initial cash flow of minus $1,000,000) and a projected life of 5 years. Estimated cash flows into the system over the 5-year period are as follows:

First year	$525,000
Second year	300,000
Third year	280,000
Fourth year	200,000
Fifth year	125,000

The following table shows that the initial $1,000,000 investment will be returned between the second and third years or to be more precise, 2.6 years by interpolation. Thus the payout time in this example is 2.6 years.

Time, end of year	Cumulative cash flow ($)
0	−1,000,000
1	−475,000
2	−175,000
3	105,000

The obvious shortcoming of a payout time determination is that it does not consider the time value of money, interest. Thus some analysts include interest as follows:

(1) End year	(2) Investment for year	(3) Interest on investment, 10% of (1)	(4) Cash flow	(5) Cash flow after interest charge, (4)–(3)	(6) Cumulative net cash flow
0			$–1,000,000		$–1,000,000
1	$1,000,000	100,000	525,000	425,000	–575,000
2	575,000	57,500	300,000	242,500	–332,500
3	332,500	33,250	280,000	246,750	–85,750
4	85,750	8,575	200,000	191,425	105,675

Including interest, the payout time increases in this example from 2.6 years to between 3 and 4 years (3.4 years by interpolation).

Although payout time with interest is probably more realistic, payout time without interest is more commonly used due to its simplicity of calculation. As a rough rule of thumb, it can be said that a potential investment is worthy of further consideration if the payout time (without interest) is 4–5 years or less.

The third concept, *return on investment* (ROI), is perhaps a more realistic measure of profitability than payout time because ROI expresses profitability as a percentage return per year. Mathematically, ROI is

$$\text{ROI (\%)} = \frac{\text{average yearly profit}}{\text{original fixed investment}} \times 100$$

Again using the foregoing example and assuming straight-line depreciation for the 5-year life, ROI is determined as follows:

(1) Year	(2) Cash flow	(3) Depreciation	(4) Profit (2)–(3)
1	$525,000	$200,000	$325,000
2	300,000	200,000	100,000
3	280,000	200,000	80,000
4	200,000	200,000	0
5	125,000	200,000	–75,000
		Total:	$430,000
		Average:	$86,000

$$\text{ROI} = \frac{\$86{,}000}{\$1{,}000{,}000} \times 100 = 8.6\% \text{ year}$$

It should be noted that some analysts calculate ROI on cash flow or gross profit before taxes rather than on net profit. Also, some people exclude working capital from the investment figure. However, since ROI is only a method of comparing alternatives, the particular variant which is used is not important *provided that* the same variant is consistently used for all comparisons.

The fourth concept, return on average investment (ROAI) is similar to ROI except that the divisor is the average outstanding investment, i.e., the investment less previously accumulated depreciation. Using the same example, the calculation is as follows:

Year		Investment
1		$1,000,000
2	$1,000,000–200,000 =	800,000
3	800,000–200,000 =	600,000
4	600,000–200,000 =	400,000
5	400,000–200,000 =	200,000
	Total:	$3,000,000
	Average:	$ 600,000

$$\text{ROAI} = \frac{\$86{,}000}{\$600{,}000} \times 100 = 14.3\% \text{ year}$$

As was true with ROI, many variants exist in the method of calculating ROAI. The variants again are of no major consequence if the same technique is used consistently for all comparisons.

The discounted cash flow (DCF) or interest-rate-of-return method is the most widely used of all profitability measures and is probably the most valid technique for assessing true profitability. The reason is that DCF considers all cash flows into and out of a project as well as time value of future cash flows. Clearly a dollar receipt at some time in the future is not worth as much as a current dollar receipt which is available for reinvestment.

In effect, the DCF calculation asks, "If I invest money today and receive back dollars in cash flow over the life of my investment, what rate of return will I receive on my investment?"

The solution is one of trial and error for the following equation:

$$0 = \sum_{0}^{n} \frac{CF_n}{(1+i)^n}$$

where

CF_n = cash flow in year n

i = interest rate of return

The equation reduces to

$$0 = -I + \sum_{1}^{n} \frac{CF_n}{(1+i)^n} \tag{12.1}$$

where

I = initial capital investment (year 0)

Further, for cases in which cash flow is constant each year over the life of the project, the equation becomes

$$I = R\left[\frac{(1+i)^n - 1}{i(1+i)^n}\right]$$

where

R = annual cash flow

n = project life, years

These equations appear to be quite complex, particularly when a trial-and-error solution is required to determine i (the DCF or interest rate of return). The solutions, however, are greatly simplified by the use of compound interest tables such as those in Appendix C. These tables tabulate the compound interest functions needed for solution of the DCF equations at various levels of interest and by time. Appendix C gives values for the following four functions:

Compound interest factor $(1 + i)^n$

Present worth factor $\dfrac{1}{(1 + i)^n}$

Unacost (uniform annual cost) present worth factor $\dfrac{(1 + i)^n - 1}{i(1 + i)^n}$

Capital recovery factor $\dfrac{i(1 + i)^n}{(1 + i)^n - 1}$

Again using the foregoing example, since cash flows differ each year, Equation 12.1 is used to calculate DCF as follows:

$$0 = -1{,}000{,}000 + \frac{525{,}000}{(1 + i)^1} + \frac{300{,}000}{(1 + i)^2} + \frac{280{,}000}{(1 + i)^3} + \frac{200{,}000}{(1 + i)^4} + \frac{125{,}000}{(1 + i)^5}$$

At $i = 20\%$, using values of the present worth factor from Appendix C, the equation yields –\$45,449.
At $i = 15\%$, the equation yields +\$43,980.

Interpolating between these two values, the equation becomes equal to zero at $i = 17.5\%$. Therefore, the DCF rate of return, in this example, is 17.5% per year.

It should be noted that the DCF calculation assumes that all cash flows during a year are credited at year end. A more precise solution would obviously be obtained by discounting on a daily or monthly basis. However, in virtually all cases the resultant answers change only slightly. This minor inaccuracy is insignificant in light of the inherent inaccuracies on the original operating and capital cost estimates themselves. Thus there is rarely any justification for daily or monthly discounting to determine DCF for a preliminary (or even a definitive) estimate.

The final technique is the *venture worth* (VW) method, which is sometimes called the incremental present worth method. It is a variation of the DCF method but avoids a trial-and-error computation by discounting at a fixed rate, usually the minimum acceptable return on capital as established by company policy. The equation is:

$$VW = -I + \sum_{1}^{n} \frac{CF_n}{(1+i)^n} \qquad (12.2)$$

Venture worth (VW) thus is the present worth of all cash flows at the specified return rate.

Again using the previous example, and assuming a minimum acceptable return of 10% per year:

$$VW = -1{,}000{,}000 + \frac{525{,}000}{(1.10)^1} + \frac{300{,}000}{(1.10)^2} + \frac{280{,}000}{(1.10)^3} + \frac{200{,}000}{(1.10)^4} + \frac{125{,}000}{(1.10)^5}$$

Using the present worth factors from Appendix C at 10%,

$$\begin{aligned} VW &= -1{,}000{,}000 + 477{,}272 + 247{,}935 + 210{,}367 + 136{,}602 + 77{,}615 \\ &= \$149{,}791 \end{aligned}$$

Thus, at a 10% return rate, the present worth of the investment and future cash flows is $149,791.

In itself this measure of profitability has absolutely no meaning. It is useful, however, in comparing alternative investments of the same magnitude.

REVENUE REQUIREMENTS

Although special techniques are usually superfluous in financial analysis, one method, revenue requirement, deserves some mention because of the extent of its use in the electric utility industry. This discussion will not pit discounted cash flow against revenue requirement. Rather it takes the position that they are complementary calculations and yield answers that can be reconciled with one another. A closely related technique to revenue requirements is the "levelized fixed charge" which is popular because it serves as a shortcut for utility engineers who need to make numerous comparisons of design alternatives.

Although basically sound, this technique is difficult to apply to major projects because of the inability to cope with interest during construction, varying project lives, and investment tax credit. Discounted cash flow is not limited by any of these conditions and provides the actual cash flow of the project on a year-to-year basis. Therefore, it is recommended that DCF, not revenue requirement, be used for all projects.

PROBLEMS AND QUESTIONS

12.1 You have made an initial investment of $1,500,000. There are no salvage or land values. Cash flows over the life of the equipment are as follows:

Year	Cash flow
1	$250,000
2	500,000
3	500,000
4	750,000
5	900,000

Will this investment be acceptable if your minimum acceptable rate of return is 25%?

12.2 Four different building locations are being evaluated for a new, high-technology plant.

MARR (Minimum Attractive Rate of Return) = 10%.

Alternatives:	A	B	C	D
Building cost ($):	200,000	275,000	190,000	350,000
Annual cash flow ($):	22,000	35,000	19,500	42,000

Which location should be chosen?

12.3 A machine is being considered for replacement. The old machine cost $300,000 and now is worth only $90,000, but it can be used for another 5 years. A new machine can be purchased for $360,000 would last 5 years, and would save $70,000 in operating costs. Should it be purchased? Compare on the basis of annual average costs assuming straight line depreciation.

12.4 Assuming an investment of $5,000 and a constant cash flow of $2,250 per year for 5 years, calculate the discounted cash flow rate of return on the investment, the payback time, the return on investment, and the venture worth at 20%. Show all calculations.

12.5 In determining cash flow of a project, what items are included? Which of these items are negative and which are positive?

12.6 An investment of $1,500,000 over a 5–year period has the following estimated cash flows:

Year 1	$450,000
Year 2	675,000
Year 3	570,000
Year 4	270,000
Year 5	180,000

Calculate the discounted cash flow rate of return on the investment.

13

Life-Cycle Costing and Sensitivity Analysis*

INTRODUCTION

Contractors, owners, and design professionals need practical methods and guidelines for evaluating the economic performance of projects. Life-cycle cost (LCC) analysis is one method for making such evaluations. LCC analysis is used to evaluate alternative systems which compete on the basis of cost. Thus only candidate systems which satisfy all performance requirements (e.g., code, safety, comfort, and reliability requirements) can be legitimately compared using the LCC method. The system alternative with the lowest LCC over the project study period is the most cost-effective choice.

An example of an appropriate use of the LCC method is choosing between an oil furnace and gas furnace for heating a building. To be candi-

---

* Adapted from "Life-Cycle Costing" by Harold E. Marshall and Stephen R. Petersen, National Institute of Standards and Technology, Gaithersburg, Maryland, which appears in *Gladstone's Mechanical Estimating Guidebook*, 6th ed., K.K. Humphreys and J. Gladstone, eds., McGraw-Hill, New York, 1995.

date projects for an LCC analysis, each system must meet the minimum performance requirements for occupant comfort, and the fuel required for each system must be available at the site. The cost-effective choice will depend on which system has lower life-cycle costs over the project study period, assuming other things equal.

Systems with the lowest LCC often have higher initial costs than competing systems. This leads contractors to ask why they should bid a higher price and put out more money up front than is necessary, particularly when they are bidding a project on a competitive basis. The answer is that the more cost-effective system can be expected to save the owner/operator of the facility enough money over time to more than pay the contractor for the increase in initial cost, thereby providing the contractor more profit.

This chapter describes how to measure life-cycle costs and use the LCC method in choosing among alternative systems. A sample problem using the LCC method shows how to find the cost-effective efficiency level of an air conditioning system. And a sample problem using the savings-to-investment ratio (SIR), an evaluation method related to LCC, shows how to find the optimal combination of independent energy conservation investments when there is too little money to do all the projects which return positive net savings.

The chapter also provides a brief overview of other measures of project worth, such as net savings and payback, and identifies circumstances when they might be appropriate. It has a section on how to treat uncertainty with sensitivity analysis. Appendix A provides definitions of the economic terms used in this chapter for your help whenever you see a term that is unfamiliar.

HOW TO MEASURE AND APPLY LCC

The LCC of a project alternative is the sum of its initial investment costs (*I*), present value of replacement costs (*R*), present value of energy costs (*E*), and the present value of operation, maintenance, and repair (*OM&R*) costs, minus the present value of salvage (*S*), which is sometimes referred to as resale value or residual value. Expressed as an equation, it is:

$$\text{LCC} = I + R - E + OM\&R - S$$

An alternative formulation that shows mathematically the discounting of future costs to present values is:

$$LCC = \sum_{t=0}^{N} \frac{C_t}{(1+d)^t}$$

where

C_t = sum of all relevant costs, less any positive cash flows such as salvage, occurring in time period t

N = number of time periods in the study period

d = investor's discount rate for adjusting cash flows to present value

Note that this equation is equivalent to venture worth (Equation 12.2) discussed in Chapter 12.

Steps in LCC Analysis

The list below shows the steps to follow in conducting an LCC analysis.

1. Identify acceptable alternatives
2. Establish common assumptions (e.g., study period, discount rate, and base date)
3. Estimate all project costs and their timing
4. Discount future costs to present value
5. Compute total LCC for each alternative
6. Identify alternative with lowest LCC
7. Consider unquantifiable costs and benefits
8. Consider uncertainty in input values
9. Compute supplementary measures of relative economic performance (if necessary)
10. Select best alternative

Rules in Applying the LCC Method

To use the LCC method, you need to compute the LCC of a project alternative (sometimes called the base case) against which to compare the LCCs of your proposed design alternative(s). Usually the base case is the alternative with the lowest initial cost.

For each alternative you consider, you must use the same study period; i.e., time over which you compare the alternatives. Only then can you determine which is more cost effective. To come up with a common study period, you will sometimes have to include replacements in short-lived projects and account for salvage value in long-lived projects. You must also use the same discount rate for each alternative.

An implicit assumption in LCC analysis is that performance or benefits from each project alternative are equal. If they are not, you must net those differences out of costs or compare those differences directly to make a valid comparison with the LCC method. If your alternatives have substantial differences in performance, use net benefits or related methods that account explicitly for benefits as well as costs.

Example: LCC Applied to the Choice of an Air Conditioning System

Suppose you are selecting a new central air conditioner for installation in a house with a design cooling load of 36,000 Btu per hour (Btuh) in a region with approximately 1,500 full-load cooling hours per year. The system with the lowest initial cost that meets the US Department of Energy's current energy performance standards has a Seasonal Energy Efficiency Ratio (SEER) of approximately 10 Btuh/W. Because the cooling load hours are above average, you will probably also want to consider systems with SEERs of 12 and 14 Btuh/W, even though their initial costs are higher. The LCC method helps you determine which SEER will result in the lowest LCC over a 15 year study period.

Local electricity rates are currently \$0.08/kWh (summer rates), with no demand charge, and are expected to increase at about 3% per year. Assume an 8% discount rate to convert future costs (including price increases) to present value. All three systems have an expected life of 15 years and approximately the same maintenance costs.

Table 13.1 shows the calculation of LCC for each alternative system, based on the sum of the initial cost and present value energy costs for each system. System B has the lowest LCC (Col. 7) and is therefore the economic choice, assuming that its reliability, maintenance, and sound characteristics are not worse than systems A or C.

Note that if the local utility were to offer a cash rebate for selecting a higher efficiency air conditioner, the initial investment cost (Col. 8) should be reduced accordingly for systems B and C. Based on the rebates reflected in Col. 8, the SEER 14 of system C becomes the most economic choice.

Typical LCC Applications

Accept/reject decisions occur when you have to decide whether or not to do a project. Examples are retrofits to plants to improve energy efficiency. These can be evaluated as independent projects as long as an investment in one does not affect the savings of another. This would be true, for example, if they were in different buildings. The LCC can help you decide if it pays to undertake any one of these projects compared to its base case. Accept a high efficiency motor, for example, if its LCC is less than the LCC of the base case.

Efficiency levels or size decisions occur when you must decide how much of something to invest in. Examples are choosing the lagging thickness on a steam line, the collector area for a solar energy system, or the level of thermal resistance for wall insulation. Make these types of decisions by choosing the level or size with the minimum LCC.

System or design decisions occur when you choose the most cost-effective of multiple alternatives for meeting an objective. For example, the LCC method can help you choose among oil or gas as a fuel; among fiber glass, foam, or cellulose for insulation material; and between types of pumps or motors. Choose the alternative with the minimum LCC as long as it satisfies system performance requirements.

OTHER METHODS AND WHEN TO USE THEM

Net Savings (NS) or Net Benefits (NB)

The NS method measures the amount of present value net savings earned over the study period from investing in a candidate project as compared to a base-case alternative. When an investment affects costs exclusively, a common approach to computing NS is to compare the LCC associated with the investment to the LCC of the base-case alternative. The NS for a heat pump, for example, is the difference between the LCC of heating and cooling an office building to a target comfort level with the base-case system, say an oil furnace with electric air conditioner, and the LCC of heating and cooling with the heat pump. If NS with the heat pump is positive, the investment makes money and is considered cost effective; if NS is zero, the investment neither makes nor loses money; if NS is negative, the investment loses money.

Use the NB method when positive revenues or benefits rather than savings accrue from the project. Either NS or NB is appropriate for accept/re-

Table 13.1 Computation of LCC for Three Air Conditioners

(1)	(2)	(3)	(4)	(5)	(6) Without utility rebate	(7)	(8) With utility rebate	(9)
System	SEER Btuh/W	Annual kWh use	Annual kWh cost, $	PV kWh cost, $	Initial AC cost, $	Total LCC, $	Initial AC cost, $	Total LCC, $
(A)	10.0	5,400	432	4,527	2,000	6,527	2,000	6,527
(B)	12.0	4,500	360	3,773	2,500	6,273	2,200	5,973
(C)	14.0	3,857	309	3,234	3,100	6,334	2,500	5,734

Explanation and computation of values

(1) alternatives
(2) from product literature
(3) = 36,000 Btuh/SEER (Btuh/W) × 1500 hr/yr
(4) = (3) × \$0.08/kWh
(5) = (4) × 10.48, where 10.48 is the UPV^* factor for an annually recurring cost increasing at a rate of 3% and discounted at 8% per year, based on the equation:

$$UPV^* = \frac{(1+e)}{(d-e)}\left[1 - \frac{(1-e)^n}{(1+d)^n}\right] \text{ when } e \neq d, \text{ else } UPV^* = n$$

where

e = the annual rate of price increase
d = discount rate, and
n = number of time periods

(6) estimated installed cost
(7) = (5) + (6)
(8) assumed rebate for SEER 12 = \$300; assumed rebate for SEER 14 = \$600
(9) = (5) + (8)

ject, size, or design decisions. For detailed information on how to calculate, apply, and interpret the NS and NB methods, see Ruegg and Marshall (1).

Savings-to-Investment Ratio (SIR)

The SIR of a project alternative is the sum of its present value savings divided by the sum of the present value of all investment-related costs attributable to that alternative over the base case:

$$SIR_{A:BC} = \frac{\sum_{t=1}^{N} S_t / (1+d)^t}{\sum_{t=0}^{N} I*_t / (1+d)^t}$$

where

$SIR_{A:BC}$ = the ratio of *PV* savings to additional *PV* investment costs of the alternative (A) relative to the base case (BC),

S_t = *PV* savings in year t in operating-related costs attributed to the alternative

$I*_t$ = additional *PV* investment-related costs attributable to the alternative in year t.

Steps in SIR Analysis

The steps in performing an SIR analysis are similar to those shown for LCC analysis, except that for the SIR you need to calculate specific savings from each alternative rather than just its LCC.

Rules in Applying the SIR Method

The main rule to remember in using the SIR is to place in the denominator of the SIR formula any investment-related cost on which you are seeking to measure your investment return. The placement of items in the numerator and denominator can alter the relative priority of a project within a group of cost-effective projects, but will not cause a project alternative with an SIR > 1.0 to fall below 1.0 or vice-versa.

Typical Applications of SIR

One use of the SIR is to make accept/reject decisions for single projects. Accept the project if the SIR > 1.0; reject it if the SIR < 1.0.

A second use of the SIR is to choose among cost-effective projects when the investment budget is limited. When money is not available to fund every system improvement that is cost effective, try to allocate funds so that the overall net savings (total savings less total investment cost) are maximized from the project selection. An appropriate method for allocating limited funds among independent projects is to rank projects in declining order of their SIRs, and then fund them in that order until the budget is exhausted. The package of investment options selected by SIR ranking will generate the greatest net savings overall relative to any other package. This works for allocating a budget within any facility or among many facilities to which a single budget applies. The projects must be functionally independent, however, for the SIR allocation method to work. That is, the savings from one project cannot significantly affect the savings from another.

The SIR ranking holds only for the additional investment required by system options which are expected to reduce future costs relative to the basic system. It is assumed that, at a minimum, the basic system must be installed for the project or process to function properly. For example, an SIR would not be calculated for the lowest initial cost (and presumably least efficient system) that meets the requirements of a new project. But it would be calculated for the additional investment required to install a higher efficiency system. Furthermore, this SIR would only be compared to SIRs for other independent improvements to determine how to allocate a fixed budget among such improvements.

Example: SIR Applied to Choosing Among Independent Projects When the Budget Is Limited

Table 13.2 shows how to use the SIR in choosing among five building improvements for energy reduction in a new building. Each improvement is cost effective because the SIR of each exceeds 1.0. But total investment costs for the improvements are $10,000, while only $5,000 is available to invest. In this case, the SIR ranking points to improvements A, B, and D. (Improvement C, while having a higher SIR than D, will break the budget, so you skip over it and choose D.) If $7,500 of funding were available, select improvements A, B, C, and D.

Table 13.2 SIR Ranking of Building Improvements

		Initial cost, $	PV savings, $	SIR
A.	Improved lighting fixtures	2,500	10,000	5.0
B.	Improved roof deck insulation	2,000	7,000	3.5
C.	Higher efficiency window systems	2,500	7,500	3.0
D.	Higher efficiency water heater	500	1,250	2.5
E.	Automatic entry doors (to reduce uncontrolled air leakage)	2,500	5,000	2.0

Payback

Payback (PB) measures how long it takes to recover investment costs. If you ignore the time value of money when you compute the payback period (i.e., you use a zero discount rate), you have simple payback (SPB). If you account for the time value of money by using a positive discount rate, you have a discounted payback (DPB). The DPB is a more accurate measure of payback because the time value of money is taken into account.

Investors sometimes specify a maximum acceptable payback period (MAPP) for evaluating a system. For example, a high efficiency heat pump might be chosen over a conventional heat pump only if the higher first cost is returned through fuel savings in, say, two years (the MAPP). But we discourage using some arbitrary MAPP to select systems because it is an unreliable guide to cost-effective choices.

PB is easy to understand, however, and as a supplementary method, it may be helpful in screening potential projects quickly. It is also useful to determine how long it takes for a project to break even. But it is a poor measure of a project's profitability over the long run because it ignores cash flows after payback. And if you use SPB, it will not even give you a reliable measure of the time to break even. So try to use LCC, NS, and SIR methods instead of PB when making economic decisions.

UNCERTAINTY AND SENSITIVITY ANALYSIS

Long-lived investments such as process plants are characterized by uncertainties regarding project life, operation and maintenance costs, and other factors that affect project economics. Since values of these variable factors are generally unknown, it is difficult to make economic evaluations with a high degree of certainty. One approach to this uncertainty problem is to use sensitivity analysis. It is a simple, inexpensive technique for handling uncertainty that gives you some perspective of how far off your measures of project worth might be.

Sensitivity analysis measures the impact on project outcomes of changing one or more key input values about which there is uncertainty. For example, a pessimistic, expected, and optimistic value might be chosen for an uncertain variable. Then an analysis could be performed to see how the outcome changes as you consider each of the three chosen values in turn, with other things held the same. When computing measures of project worth, for example, sensitivity analysis shows you just how sensitive the economic payoff is to uncertain values of a critical input, such as the discount rate or project maintenance costs expected to be incurred over the project's study period. Sensitivity analysis reveals how profitable or unprofitable the project might be if input values to the analysis turn out to be different from what is assumed in a single-answer approach to measuring project worth.

Sensitivity analysis can also be performed on different combinations of input values. That is, several variables are altered at once and then a measure of worth is computed. For example, one scenario might include a combination of all pessimistic values, another all expected values, and a third all optimistic values. Note, however, that sensitivity analysis can in fact be misleading (2) if all pessimistic assumptions or all optimistic assumptions are combined in calculating economic measures. Such combinations of inputs would be unlikely in the real world.

Sensitivity analysis can be performed for any measure of worth. And since it is easy to use and understand, it is widely used in the economic evaluation of government and private sector projects. The Office of Management and Budget (3) recommends sensitivity analysis to Federal agencies as one technique for treating uncertainty in input variables. And the American Society for Testing and Materials, in its *Standard Guide for Selecting Techniques for Treating Uncertainty and Risk in the Economic Evaluation of Buildings and Buildings Systems* (4), describes sensitivity analysis for use in government and private sector applications.

Sensitivity Analysis Application

The results of sensitivity analysis can be presented in text, tables, or graphs. The following illustration of sensitivity analysis applied to a programmable control system uses text and a simple table.

Consider a decision on whether or not to install a programmable time clock to control heating, ventilating, and air conditioning (HVAC) equipment in a commercial building. The time clock would reduce electricity consumption by turning off that part of the HVAC equipment that is not needed during hours when the building is unoccupied.

Using net savings (NS) as the measure of project worth, the time clock is acceptable on economic grounds if its NS is positive, that is, if its present value savings exceed present value costs. The control system purchase and maintenance costs are felt to be relatively certain. The savings from energy reductions resulting from the time clock, however, are not certain. They are a function of three factors: the initial price of energy, the rate of change in energy prices over the life cycle of the time clock, and the number of kilowatt hours (kWh) saved. Two of these, the initial price of energy and the number of kWh saved, are relatively certain. But future energy prices are not. To test the sensitivity of NS to possible energy price changes, three values of energy price change are considered: a low rate of energy price escalation (slowly increasing benefits from energy savings); a moderate rate of escalation (moderately increasing benefits); and a high rate of escalation (rapidly increasing benefits).

Table 13.3 shows three NS estimates that result from repeating the NS computation for each of the three energy price escalation rates. To appreciate the significance of these findings, it is helpful to consider what extra information you now have over the conventional single-answer approach where,

Table 13.3 Energy Price Escalation Rates

Energy price escalation rate	Net savings ($)
Low	–15,000
Moderate	20,000
High	50,000

say, you computed a single NS estimate of $20,000. The table shows that the project could return up to $50,000 in NS if future energy prices escalated at a high rate. On the other hand, you see that the project could lose as much as $15,000. This is considerably less than breakeven, where the project would at least pay for itself. It is also $35,000 less than what you calculated with the single-answer approach. Thus sensitivity analysis alerts you that accepting the time clock could lead to an uneconomic outcome.

There is no explicit measure of the likelihood that any one of the NS outcomes will happen. The analysis simply tells you what the outcomes will be under alternative conditions. However, if there is reason to expect energy prices to rise, at least at a moderate rate, then the project very likely will make money, other factors remaining the same. This adds helpful information over the traditional, single-answer approach to measures of project worth.

Advantages and Disadvantages

There are several advantages of using sensitivity analysis in cost engineering studies. First, it shows how significant any given input variable is in determining a project's economic worth. It does this by displaying the range of possible project outcomes for a range of input values. This shows decision makers the input values that would make the project a loser or winner. Sensitivity analysis also helps you identify critical inputs so that you can choose where to spend extra resources in data collection and in improving data estimates.

Second, sensitivity analysis is an excellent technique to help you anticipate and prepare for the "what if" questions that you will be asked when presenting and defending a project. When you are asked what the outcome will be if operating costs are fifty percent more expensive than you think, you will be ready with an answer. Generating answers to "what if" questions will help you assess how well your proposal will stand up to scrutiny.

Third, sensitivity analysis does not require the use of probabilities as do many techniques for treating uncertainty.

Fourth, you can use sensitivity analysis on any measure of project worth.

And finally, you can use it when there is little information, resources, and time for more sophisticated techniques.

The major disadvantage of sensitivity analysis is that there is no explicit probabilistic measure of risk exposure. That is, while you might be sure that one of several outcomes might happen, the analysis contains no explicit measure of their respective likelihoods.

PROBLEMS*

13.1 How much is it worth to you today to avoid a cost of $5,000 in 8 years if your discount rate is 6%?

13.2 Suppose you work for an organization whose minimum acceptable rate of return (the discount rate) is 12%. Would you recommend acceptance of a project which costs $200,000 up front and will save $20,000 a year for the next 20 years?

13.3 You need to purchase euqipment which requires maintenance and which must last for 10 years. You have two choices:

a) You can buy the equipment from Supplier A who charges $8000 more in initial cost for the equipment than Supplier B, but includes maintenance at no additional charge over the life of the equipment.

b) You can buy the equipment from Supplier B and purchase a separate annual maintenance contract from Supplier B. The annual charge is $1,000 as of now, but the contract includes an escalation clause that increases the amount by 2% each year. Payment is due at the end of each year, with the first being due one year from now. Your company's discount rate is 12%.

Which supplier do you choose? Support your conclusion with appropriate calculations.

13.4 Would you recommend adding storm windows over existing windows in an office building in Boston instead of replacing them with new double-glazed windows on grounds of cost effectiveness. You plan to sell the building in 12 years. Pertinent data is:

	Storm windows	Double glazing
Window costs	$3150	$10,500
Painting costs every 5 years	758	0
Additional cleaning costs/year	90	0
Additional resale value	40% of storm window cost	60% of double glazing cost

* These problems are taken from *Introduction to Life-Cycle Costing: Video Training Workbook* by Rosalie T. Ruegg, NISTIR 4309, U.S. Department of Commerce, National Institute of Standards and Technology, Gaithersburg, Maryland, April 1990.

Discount rate %	12	12
Inflation ratc, %	5	5

In addition to this information, what other factors might be important to the decision?

REFERENCES

1. Ruegg, R.T. and H.E. Marshall, *Building Economics: Theory and Practice,* Chapman and Hall, New York, August 1990.
2. Hillier, F., "The Dcrivation of Probabilistic Information for the Evaluation of Risky Investments," *Management Science*, 1969, p. 444.
3. Office of Management and Budget, "Guidelines and Discount Rates for Benefit-Cost Analysis of Federal Programs," Circular A-94, pp. 12–13 (rev. 29 October 1992).
4. American Society for Testing and Materials, "Standard Guide for Selecting Techniques for Treating Uncertainty and Risk in the Economic Evaluation of Buildings and Building Systems," E1369–93, *ASTM Standards on Buildings Economics*, 3rd ed., Philadelphia, PA, 1994.

14

The Finished Report

The final cost report takes whatever form that fits the anticipated audience. Do not write a 97-page report with the summary and conclusions on page 97 for the senior executives of the company; they will never get to page 97. The essence of the report should be up front condensed to two pages (or less if possible) with backup material, etc., following. This gives the busy executive the summary and conclusion up front and supplies backup information if he or she cares to read further.

The capital estimate must be completed before the operating cost estimate because much of the necessary operating cost information comes from the back-up calculations developed for the capital estimate.

The finished report summary contains the pertinent information from the capital estimate, the annual operating cost estimate, and the conclusions from the financial analysis. The summary also should contain a title and a sentence or two concerning the purpose of the plant or process and what the products, if any, are.

The format of a cost report depends upon how the boss wants the results presented. However, there are several items that should be included in the report and are rather basic to getting the pertinent information incorporated into an understandable document.

As implied earlier, the summary and conclusion should be the first entity in a cost report because many company executives are interested in this section only. The summary and conclusions should include the following items:

A one-paragraph description of what the system does with the major raw material to produce the major product
The total estimated capital investment
The total estimated annual operating costs at full production
A tabular summary of the financial analysis—normally a discounted cash flow rate of return (DCF)
A tabular presentation of the significant sensitivity measurements such as the effect of raw material costs and product prices on the DCF.

The summary and conclusions should be followed by at least a generic process description. The discussion should include all of the major processing and support systems which make up the complex. Writing a description organized in the same sequence as the capital estimate affords an additional opportunity to check the logic of the processing and to reassure that omissions have not been made. It is important that the report include process flowsheets, so reference can be made while reading the report. How detailed the description should be depends upon whether or not additional documents are available on the process. If the cost report is a stand-alone document, the process description should be relatively complete but not detailed. The description should not discuss control loops, types of valves, etc. This type of information should be in a separate document.

Following the process description, the capital cost determination should be discussed in more detail than afforded in the summary and conclusions. The tabular presentation, normally the first table in a cost report, should be referred to and explained in the writeup. Also, any other tables supporting the capital estimate should be discussed (such as capital summaries for each plant processing and support system, working capital, and contingency determinations). The detailed equipment lists incorporated in most cost reports or in another document for large reports should be referenced for the reader. If major assumptions have been made to facilitate completing the estimate, a table listing same should be included in the report and discussed in the capital estimate writeup.

The operating cost estimate cannot be completed until after the capital estimate is finished; thus, the operating cost determination should be discussed next. As implied earlier, almost all of the information required for the operating cost estimate is calculated either for the capital estimate or to support the same. The basic documents used in development of the operating costs should be referenced and included in the cost report, i.e., the materials balance (overall flowsheet), the utility balance, the staffing plan or basis for labor, and the local taxes and insurance determination, whether generic or actual. The operating cost discussion should give the reader an understanding of the completeness of the documents upon which the operating cost estimate has been based.

Rarely is a plant started at full capacity, but rather goes through a startup period. Depending on the complexity of the plant and the newness of the technology involved, the startup period may require as much as three years operation at less than capacity. The estimators and engineers must estimate the length of the startup period as well as the estimated production of specification product for each startup year. These projections are difficult to make; however, such estimates are required to prevent surprises for management and to instill reality in the financial analysis. An estimated operating cost must be determined for each startup year. These estimates should be included in the cost report along with the basis for determining same.

The financial analysis, usually a discounted cash flow rate of return (DCF), is discussed next. DCF may be defined as the discount rate at which the present values of all of the cash flows in and out of the venture accumulate to zero over the life of the project. A rate, normally called a hurdle rate or minimum acceptable rate of return (MARR), is set by most companies and is the lowest DCF that the company will accept on a project. This hurdle rate for many companies is about 20% based on project economics. Project economics implies a DCF based on 100% equity financing with consideration for anticipated inflation through the construction and operating periods. The definition may be modified from the above, but all projects should be compared using the same financial basis. Otherwise, the comparisons are invalid. Keep in mind that a bad project can be made to appear to be a good investment by the manipulation of numbers such as low capital estimating, leveraged financing, low interest rates, inflated product prices, etc. Also, a good project may be made to appear to be a poor investment by overly conservative assumptions and projections. The key to financial analysis is that *objectivity* must be maintained. That is the goal of a financial analysis—to get the correct answer, not necessarily the most palatable one.

To initiate a DCF analysis, a number of chores must have been accomplished, including determining a capital expenditure schedule, projecting a rate of inflation, estimating annual operating cost during startup and at full capacity, and either establishing a product price or prices to determine DCF or using a DCF rate to determine the major product price. In the latter case, prices for coproducts will have to be estimated or mathematically related to the major product price.

The capital expenditure schedule can be estimated early in project development but is more definitive as the project progresses, much the same as other cost numbers. The distribution of the capital costs early in a project is much more important than those capital costs distributed near the end of the project because of the time value of money and/or the discount rate. The present value of a dollar spent today is one dollar, but the present value of a dollar discounted at 15% is 0.061 dollars in the twentieth year.

It is good practice to include in the cost report the backup material for each line item that appears on the capital cost summary table. For example, if one line item is raw material receiving and storage, then the backup summary for the line item should be included. The backup should also be included for offsite and support systems.

If the report is preliminary and major assumptions have been made to facilitate completion of the estimate, then a table listing such assumptions should also be included.

14.1 FINAL PROBLEM—A COMPLETE COST ESTIMATE

You are asked by your supervisor to prepare a preliminary cost estimate on a proposed plant addition to determine its profitability. The available information is as follows:

Increase in production rate: 700 tons per week at 100% of capacity.
The plant operates 3 shifts/day, 7 days/week.
Product reject rate is 1%.

Installed process equipment cost	$705,220
Mobile equipment (trucks, etc.) cost	22,880
Buildings cost	400,000
Land cost	200,000
Working capital required	189,030
Total new capital investment	$1,517,130

Instrumentation is assumed at 5% of process equipment cost; engineering at 10% of process equipment cost, and contingencies at 5% of process equipment cost.

Raw materials costs are:

Material A	$1.00/ton
Material B	$1.00/ton
Material C	$117.63/ton
Process water	$0.10/1000 gal

The product composition is:

72.0 wt% Material A (dry basis)
25.2 wt% Material B (dry basis)
2.8 wt% Material C (dry basis)
5.0 wt% water

Reject product is recycled to the process and substituted for a portion of the Material B requirements.

Anticipated plant downtime is 10%.

Utility costs, per ton of saleable product for the plant addition are:

Electric power	$0.54
Natural gas	$2.18

The addition will require one new production worker at $21.00/hr.

First-line supervision costs are projected at 10% of production labor cost.

Maintenance cost/year is estimated at 4% of depreciable capital investment/year (1/2 material—1/2 labor).

Operating (factory) supplies are assumed to cost 1/2% of depreciable capital investment/year.

Packing and shipping costs are estimated at $2.70/ton of product.

Depreciation guidelines (straight line) are:

Process equipment	15 yr
Mobile equipment	3 yr
Buildings	30 yr

Salvage values are assumed to be 0
Equipment will be replaced as needed
Plant life is 30 years

Taxes and insurance equal 2% of depreciable capital investment/year.

All overheads (payroll, laboratory, and indirect) total 75% of total labor (production labor + maintenance labor + first-line supervision).

The market value of the product is $37.80/ton.
Semivariable costs at 0 production = 35% of semivariable costs at 100% of production.

a) What will be the total production cost (including depreciation) per ton?
b) What is the estimated gross profit/year before taxes?
c) What is the net after tax profit (state and federal taxes = 51%)?
d) What is the Discounted Cash Flow (DCF) rate of return over the life of the project?
e) What is the breakeven point (% of capacity)?
f) What is the shutdown point (% of capacity)?
g) Assuming that the minimum acceptable rate of return is 15%, at what point (% of capacity) will the desired return be achieved?

Appendixes

Appendix A

Cost Engineering Terminology*

Accounts payable The value of goods and services available for use on which payment has not yet been made. (*See also* **Taxes payable.**)

Accounts receivable The value of goods shipped or services rendered to a customer on which payment has not yet been received. Usually includes an allowance for bad debts.

Administrative expense The overhead cost due to the non-profit-specific operations of a company. Generally includes top management salaries, and the costs of legal, central purchasing, traffic, accounting, and other staff functions and their expenses for travel and accommodations.

---

* Many of the definitions in this glossary are excerpted from *Standard Cost Engineering Terminology*, AACE Standard No. 10S-90, AACE International, Morgantown, WV, 1990.

Amortization (1) As applied to a capitalized asset, the distribution of the initial cost by periodic charges to operations as in depreciation; most properly applies to assets with indefinite life; (2) the reduction of a debt by either periodic or irregular payments; (3) a plan to pay off a financial obligation according to some prearranged schedule.

Annual equivalent (1) In time value of money, a uniform annual amount for a prescribed number of years that is equivalent in value to the present worth of any sequence of financial events for a given interest rate; (2) one of a sequence of equal end-of-year payments which would have the same financial effect when interest is considered as another payment or sequence of payments which are not necessarily equal in amount or equally spaced in time. (*See also* **Average annual cost**.)

Annuity (1) An amount of money payable to a beneficiary at regular intervals for a prescribed period of time out of a fund reserved for that purpose; (2) a series of equal payments occurring at equal periods of time.

Average annual cost The conversion, by an interest rate and present worth technique, of all capital and operating costs to a series of equal payments occurring at equal periods of time.

Average interest method A method of computing required return on investment based on the average book value of the asset during its life or during a specified study period.

Base-case The base-case alternative is the alternative against which proposed alternatives are compared.

Base date The date (usually the beginning of the study period) to which benefits and costs are converted to time equivalent values when using the present value method.

Battery limit Comprises one or more geographic boundaries, imaginary or real, enclosing a plant or unit being engineered and/or erected, established for the purpose of providing a means of specifically identifying certain portions of the plant, related groups of equipment, or associated facilities. Generally refers to the processing area and includes all process equipment, and excludes such other facilities as storage, utilities, administration buildings, or auxiliary facilities.

Book value (net) (1) Current investment value on the books calculated as original value less depreciation accruals; (2) new asset value for accounting use; (3) also, the value of an outstanding share of stock of a corporation at any one time, determined by the number of shares of that class outstanding.

Breakeven chart A graphic representation of the relation between total income and total costs for various levels of production and sales indicating areas of profit and loss.

Breakeven point(s) (1) In business operations, the rate of operations, output, or sales at which income is sufficient to equal operating cost, or operating cost plus additional obligations that may be specified; (2) the operating condition, such as output, at which two alternatives are equal in economy; (3) the percentage of capacity operation of a manufacturing plant at which income will just cover expenses (4) a combination of benefits (savings or revenues) that just offset costs, such that a project generates neither profits nor losses.

Burden In construction, the cost of maintaining an office with staff other than operating personnel. Includes also federal, state, and local taxes, fringe benefits, and other union contract obligations. In manufacturing, burden sometimes denotes overhead.

Capacity factor (1) The ratio of average load to maximum capacity; (2) the ratio between average load and the total capacity of the apparatus, which is the optimum load; (3) the ratio of the average actual use to the available capacity. Also called capacity utilization factor.

Capital budgeting A systematic procedure for classifying, evaluating, and ranking proposed capital expenditures for the purpose of comparison and selection, combined with the analysis of the financing requirements.

Capital, cost of The weighted average of (a) the after-tax cost of long-term debt, (b) the yield of any outstanding preferred stock, and (c) the cost of common equity capital. Usually expressed as a percent.

Capital, direct Cost of all material and labor involved in the fabrication, installation, and erection of facilities.

Capital, fixed The total original value of physical facilities which are not carried as a current expense on the books of account and for which depreciation is allowed by the federal government. It includes plant equipment, building, furniture and fixtures, and transportation equipment used directly in the production of a product or service. It includes all costs incidental to getting the property in place and in operating condition, including legal costs, purchased patents, and paid-up licenses. Land, which is not depreciable, is often included. Characteristically, fixed capital cannot be converted readily into cash.

Capital, indirect Costs associated with construction but not directly related to fabrication, installation, and erection of facilities. Can be broken down into field costs (temporary structures, field supervision) and office costs (engineering, drafting, purchasing, and office overhead expenses).

Capital recovery (1) Charging periodically to operations amounts that will ultimately equal the amount of capital expenditure (*see* **Amortization, Depletion**, and **Depreciation**); (2) the replacement of the original cost of an asset plus interest; (3) the process of regaining the net investment in a project by means of revenue in excess of the costs from the project. (Usually implies amortization of principal plus interest on the diminishing unrecovered balance.)

Capital recovery factor A factor used to calculate the sum of money required at the end of each of a series of periods to regain the net investment of a project plus the compounded interest on the unrecovered balance.

Capital, sustaining The fixed capital requirements to (1) maintain the competitive position of a project throughout its commercial life by improving product quality, related services, safety, or economy; or (2) required to replace facilities which wear out before the end of the project life.

Capital, total Sum of fixed and working capital.

Capital, venture Capital invested in technology or markets new at least to the particular organization.

Capital, working The funds in addition to fixed capital and land investment which a company must contribute to the project (excluding startup expense) to get the project started and meet subsequent obligations as they come due. Includes inventories, cash, and accounts receivable minus accounts payable. Characteristically, these funds can be converted readily into cash. Working capital is normally assumed recovered at the end of the project.

Capitalized cost (1) The present worth of a uniform series of periodic costs that continue for an indefinitely long time (hypothetically infinite); not to be confused with a capitalized expenditure; (2) the value at the purchase date of the first life of the asset of all expenditures to be made in reference to this asset over an indefinite period of time; this cost can also be regarded as the sum of capital which, if invested in a fund earning a stipulated interest rate, will be sufficient to provide for all payments required to maintain the asset in perpetual service.

Cash costs Total cost excluding capital and depreciation spent on a regular basis over a period of time, usually 1 year. Cash costs consist of manufacturing cost and other expenses such as transportation cost, selling expense, research and development cost, or corporate administrative expense.

Cash flow The net flow of dollars into or out of the proposed project. The algebraic sum, in any time period, of all cash receipts, expenses, and investments.

Cash return, percent of total capital Ratio of average depreciation plus average profit to total fixed and working capital, for a year of capacity sales. Under certain limited conditions, this figure closely approximates that calculated by profitability index techniques where it is defined as the difference, in any time period, between revenues and all cash expenses, including taxes.

Compound amount The future worth of a sum invested (or loaned) at compound interest.

Compound amount factor (1) The function of interest rate and time that determines the compound amount from a stated initial sum; (2) a factor which when multiplied by the single sum or uniform series of payments will give the future worth at compound interest of such single sum or series.

Compound interest (1) The type of interest that is periodically added to the amount of investment (or loan) so that subsequent interest is based on the cumulative amount; (2) the interest charges under the condition that interest is charged on any previous interest earned in any time period, as well as on the principal.

Compounding, continuous (1) A compound interest situation in which the compounding period is zero and the number of periods infinitely great, a mathematical concept that is practical for dealing with frequent compounding and small interest rates; (2) a mathematical procedure for evaluating compound interest factors based on a continuous interest function rather than discrete interest periods.

Compounding period The time interval between dates at which interest is paid and added to the amount of an investment or loan. Designates frequency of compounding.

Contingencies Specific provision for unforeseeable elements of cost within the defined project scope; particularly important where previous experience relating estimates and actual costs has shown that unforeseeable events which will increase costs are likely to occur.

Contracts Legal agreement between two or more parties, which may be of the types enumerated below:

1. In *Cost-plus* contracts the contractor agrees to furnish to the client services and material at actual cost, plus an agreed-upon fee for services. This type of contract is employed most often when the scope of service to be provided is not well defined.
   a. *Cost-plus-percentage burden and fee*—the client will pay all costs as defined in the terms of the contract, plus "burden and fee" at a specified percent of the labor costs which the client is paying for directly. This type of contract generally is used for engineering services. In contracts with some governmental agencies, burden items are included in indirect cost.
   b. *Cost-plus fixed fee*—the client pays costs as defined in the contract document. Burden on reimbursable technical labor cost is considered in this case as part of cost. In addition to the costs and burden, the client also pays a fixed amount as the contractor's fee.
   c. *Cost plus fixed sum*—the client will pay costs defined by contract plus a fixed sum which will cover nonreimbursable costs and provide for a fee. This type of contract is used in lieu of a cost-plus-fixed-fee contract where the client wishes to have the contractor assume some of the risk for items which would be reimbursable under a cost-plus-fixed-fee type of contract.
   d. *Cost plus percentage fee*—the client pays all costs, plus a percentage for the use of the contractor's organization.

2. Fixed-price types of contract are ones wherein a contractor agrees to furnish services and material at a specified price, possibly with a mutually agreed upon escalation clause. This type of contract is most often employed when the scope of services to be provided is well defined.
   a. *Lump sum*—the contractor agrees to perform all services as specified by the contract for a fixed amount. A variation of this type may include a turnkey arrangement where the contractor guarantees quality, quantity, and yield on a process plant or other installation.
   b. *Unit price*—contractor will be paid at an agreed-upon unit rate for services performed. For example, technical workhours will be paid for at the unit price agreed upon. Often field work is assigned to a subcontractor by the prime contractor on a unit price basis.
   c. *Guaranteed maximum (target price)*—a contractor agrees to perform all services as defined in the contract document guaranteeing that the total cost to the client will not exceed a stipulated maximum figure. Quite often, these types of contracts will contain special share-of-the-saving arrangements to provide incentive to the contractor to minimize costs below the stipulated maximum.
   d. *Bonus-penalty*—a special contractual arrangement usually between a client and a contractor wherein the contractor is guaranteed a bonus, usually a fixed sum of money, for each day the project is completed ahead of a specified schedule and/or below a specified cost, and agrees to pay a similar penalty for each day of completion after the schedule date or over a specified cost up to a specified maximum either way. The penalty situation is sometimes referred to as liquidated damages.

Construction cost The sum of all costs, direct or indirect, inherent in converting a design plan for material and equipment into a project ready for operation, i.e., a sum of field labor, supervision, administration, tools, field office expense, and field-purchased material costs.

Cost control The application of procedures to follow the progress of design and construction projects as well as manufacturing operations in order to minimize cost with the objective of increasing profitability and assuring efficient operations. There are three essential elements of control. The first is to establish the optimum condition, the second is to measure variation from the optimum, and the third is to take corrective action in order to minimize this variation. The application of these procedures attempts to limit costs to those authorized for capital projects or cost standards, focuses control efforts where they will be most effective, and achieves maximum control at minimum operating cost.

Cost effective The condition whereby the present value benefits (savings) of an investment alternative exceed its present value costs.

Cost engineer An engineer whose judgment and experience are utilized in the application of scientific principles and techniques to problems of cost estimation; cost control; business planning and management science; profitability analysis; and project management, planning, and scheduling.

Cost index (price index) A number which relates the cost of an item at a specific time to the corresponding cost at some arbitrarily specified time in the past.

Declining balance depreciation A method of computing depreciation in which the annual charge is a fixed percentage of the depreciated book value at the beginning of the year to which the depreciation applies. Also known as percent on diminishing value.

Demand factor (1) The ratio of the maximum instantaneous production rate to the production rate for which the equipment was designed; (2) the ratio between the maximum power demand and the total connected load of the system.

Depletion (1) A form of capital recovery applicable to extractive property (e.g., mines); can be on a unit-of-output basis the same as straight-line depreciation related to original or current appraisal of extent and value of deposit (known as percentage depletion); (2) lessening of the value of an asset due to a decrease in the quantity available. It is similar to depreciation except that it refers to such natural resources as coal, oil, and timber in forests.

Depreciated book value The first cost of the capitalized asset minus the accumulation of annual depreciation cost charges.

Depreciation (1) Decline in value of a capitalized asset; (2) a form of capital recovery applicable to a property with a lifespan greater than 1 year, in which an appropriate portion of the asset's value is periodically charged to current operations.

Development costs Those costs specific to a project, either capital or expense items, which occur prior to commercial sales found necessary in determining the potential of that project for consideration and eventual promotion. Major cost areas include process, product, and market research and development.

Direct cost:

1. In construction—cost of installed equipment, material, and labor directly involved in the physical construction of the permanent facility.
2. In manufacturing, service and other nonconstruction industries—the portion of operating costs that is usually assignable to a specific product or process area. Usually included are:
   a. *Input material costs*—raw materials which appear in some form as a product. For example, water added to resin formulation is an input material, but sulfuric acid catalyst, consumed in manufacturing high-octane alkylate, is not.
   b. *Operating, supervision, and clerical payroll costs*—wages and salaries paid to personnel who operate the production facilities.
   c. *Fringe benefits*—payroll costs other than wages not paid directly to the employee. They include costs for: holidays, vacations, sick leave; Federal old age insurance; pensions, life insurance, savings plans, etc. In contracts with some governmental agencies these items are included in indirect cost.
   d. *Maintenance cost*—expense incurred to keep manufacturing facilities operational. It consists of: maintenance payroll cost; maintenance materials and supplies cost. Maintenance materials which have a life of more than 1 year are usually considered capital investment in detailed cash flow accounting.
   e. *Utilities*—the fuel, steam, air, power, and water which must be purchased or generated to support the plant operation.
   f. *Catalysts, chemicals, and operating supplies*—materials consumed in the manufacturing operations, but not appearing as a product. Operating supplies are a minor cost in process industries and are sometimes assumed to be in the maintenance materials estimate; but in many industries, mining for example, they are a significant proportion of direct cost.

g. Miscellaneous

i) *Royalties*-cost paid to others for use of a proprietary process. Both paid up and "running" royalties are used. Cost of paid-up royalties are usually on the basis of production rate. Royalties vary widely, however, and are specific for the situation under consideration.

ii) *Packaging cost*—material and labor necessary to place the product in a suitable container for shipment. Also called packaging and container cost or packing cost. Sometimes considered an indirect cost together with distribution costs such as for warehousing, loading, and transportation.

Although the direct costs described above are typical and in general use, each industry has unique costs which fall into the "direct cost" category. A few examples are equipment rental, waste disposal, contracts, etc.

Discount rate The minimum acceptable rate of return used in converting benefits and costs occurring at different times to their equivalent values at a common time. Discount rates reflect the investor's time value of money (or opportunity cost). "Real" discount rates reflect time value apart from changes in the purchasing power of the dollar (i.e., exclude inflation or deflation) and are used to discount constant dollar cash flows. "Nominal" or "market" discount rates include changes in the purchasing power of the dollar (i.e., include inflation or deflation) and are used to discount current dollar cash flow.

Discounted cash flow (1) The present worth of a sequence in time of sums of money when the sequence is considered as a flow of cash into and/or out of an economic unit; (2) an investment analysis which compares the present worth of projected receipts and disbursements occurring at designated future times in order to estimate the rate of return from the investment or project. Also called discounted cash flow rate of return, interest rate of return, internal rate of return, investor's method, or profitability index.

Discounting A procedure for converting a cash flow that occurs over time to an equivalent amount at a common time.

Distribution The broad range of activities concerned with efficient movement of finished products from the end of the production line to the consumer; in some cases it may include the movement of raw materials from the source of supply to the beginning of the production line. These activities include freight transportation, warehousing, material handling, protective packaging, inventory control, plant and warehouse site selection, order processing, market and sales forecasting, customer service, and attendant management information systems; in some cases it may include buying activities.

Economic return The profit derived from a project or business enterprise without consideration of obligations to financial contributors and claims of others based on profit.

Effective interest The true value of interest rate (effective annual interest) computed by equations for compound interest rate for a 1-year period.

Escalation The provision in actual or estimated costs for an increase in the costs of equipment, material, labor, etc. over those specified in the contract, due to continuing price level changes over time.

Estimate An evaluation of all the costs of the elements of a project or effort as defined by an agreed-upon scope. Three specific types based on degree or definition or a process industry plant are:

1. *Order of magnitude estimate*—an approximate estimate made without detailed engineering data. Some examples would be: an estimate from cost capacity curves, an estimate using scale-up [or scale-down] factors, and an approximate ratio estimate. It is normally expected that an estimate of this type would be accurate within +50% or –30%.
2. *Budget estimate*—budget in this case applies to the owner's budget and not to the budget as a project control document. A budget estimate is prepared with the use of flowsheets, layouts, and equipment details. It is normally expected that an estimate of this type would be accurate within +30% or –15%.
3. *Definitive estimate*—as the name implies, this is an estimate prepared from very defined engineering data. The engineering data include as a minimum, fairly complete plot plans and elevations, piping and instrument diagrams, one-line electrical diagrams, equipment data sheets and quotations, structural sketches, soil data and sketches of major foundations, building sketches, and a complete set of specifications. This category of estimate covers all types from the minimum described above to the maximum definitive type which

would be made from "approved for construction" drawings and specifications. It is normally expected that an estimate of this type would be accurate within +15% and –5%.

Expansion Any increase in the capacity of a plant facility or unit, usually by added investment. The scope of its possible application extends from the elimination of problem areas to the complete replacement of an existing facility with a larger one.

Expense Expenditures of short-term value, including depreciation, as opposed to land and other fixed capital.

Fee The charge for the use of a contractor's organization for the period and to the extent specified in the contract.

Field cost Engineering and construction costs associated with the construction site rather than with the home office.

Fringe benefits Employee welfare benefits; expenses of employment not paid directly to the employee, such as holidays, sick leave, supplemental unemployment benefits, Social Security, insurance, etc.

Future worth (1) The equivalent value at a designated future date based on time value of money; (2) the monetary sum, at a given future time, which is equivalent to one or more sums at given earlier times when interest is compounded at a given rate.

Home office cost Those necessary costs involved in the conduct of everyday business which can be directly assigned to specific projects, processes or end products, such as engineering, procurement, expediting, inspection, estimating, reproduction, telephone, telegraph, etc.

Hurdle rate (*see* **Required return**)

Indirect costs (1) In construction, all required costs which do not become a final part of the installation and may include, but are not limited to, field administration, direct supervision, capital tools, startup costs, contractor's fees, insurance, taxes, etc. (2) in manufacturing, costs not directly assignable to the end product or process, such as overhead and general-purpose labor, or costs of outside operations, such as transportation and distribution. Indirect manufacturing costs sometimes include insurance, property taxes, maintenance, depreciation, packaging, warehousing, and loading. In government contracts, indirect cost is often calculated as a fixed percent of direct payroll cost.

In-place value Value of a physical property, i.e., market value plus costs of transportation to site and installation.

Intangibles (1) In economy studies, conditions or economy factors that cannot be readily evaluated in quantitative terms as in money; (2) in accounting, the assets that cannot be reliably evaluated (e.g., goodwill).

Interest (1) Financial share in a project or enterprise; (2) periodic compensation for the lending of money; (3) in economy study, synonymous with required return, expected profit, or charge for use of capital; (4) the cost for the use of capital. Sometimes referred to as the time value of money.

Interest rate The ratio of the interest payment to the principal for a given unit of time and is usually expressed as a percentage of the principal.

Interest rate, compound The rate earned by money expressed as a constant percentage of the unpaid balance at the end of the previous accounting period. Typical time periods are yearly, semiannually, monthly, and instantaneous.

Interest rate, effective An interest rate for a stated period (per year unless otherwise specified) that is the equivalent of a smaller rate of interest that is more frequently compounded.

Interest rate, nominal The customary type of interest rate designation on an annual basis without consideration of compounding periods. The usual basis for computing periodic interest payments.

Interest rate of return (*see* **Profitability index**)

Inventory The value of raw materials, products in process, and finished products required for plant operation. Also other supplies, i.e., for maintenance, catalyst, chemicals, spare parts.

Investment Used interchangeably with capital, but it may include expense associated with capital outlays such as mine development.

Investor's method (*see* **Discounted cash flow**)

Labor cost, nonmanual In construction, normally refers to field personnel other than craftsmen and includes field administration and field engineering.

Labor factor The ratio between the workhours required to perform a task under project conditions and the workhours required to perform an identical task under standard conditions.

Learning curve (1) A useful planning technique particularly in the project-oriented industries where new products are phased in rather frequently. The basis for the learning curve calculation is the fact that workers will be able to produce the product more quickly after they get used to making it. (2) A graphic representation of the progress in production effectiveness as time passes.

Life (1) *Economic*: that period of time after which a machine or facility should be discarded or replaced because of its excessive costs or reduced profitability; the economic impairment may be absolute or relative; (2) *Physical*: that period of time after which a machine or facility can no longer be repaired in order to perform its design function properly; (3) *Service*: the period of time that a machine or facility will satisfactorily perform its function without major overhauls. (*See also* **Venture life**.)

Life-cycle cost (LCC) The sum of all discounted costs of acquiring, owning operating, and maintaining a project over the study period. Comparing life cycle costs among mutually exclusive projects of equal performance is one way of determining relative cost effectiveness.

Load factor (1) A ratio that applies to physical plant or equipment: average load/maximum demand, usually expressed as a percentage; equivalent to percent of capacity operation if facilities just accommodate the maximum demand; (2) is defined as the ratio of average load to maximum load.

Lot size The number of units in the lot.

Maintenance The expense, both for labor and materials, required to keep equipment or other installations in suitably operable condition. Maintenance does not usually include those items which cannot be expended within the year purchased and must be considered fixed capital.

Manufacturing cost The total of variable and fixed or direct and indirect costs chargeable to the production of a given product, usually expressed in cents or dollars per unit of production, or dollars per year. Transportation and distribution costs, and research, development, selling, and corporate administrative expenses are usually excluded.

Marginal analysis An economic concept concerned with those incremental elements of costs and revenue which are associated directly with a specific course of action, normally using available current costs and revenue as a base and usually independent of traditional accounting allocation procedures.

Marginal cost (1) the cost of one additional unit of production, activity, or service; (2) the rate of change of costs with production or output.

Marketing The broad range of activities concerned primarily with the determination of consumer or user demands or desires, both existing and potential; the satisfaction of these demands or desires through innovation or modification; and the building of buyer awareness of product or service availability through sales and advertising efforts.

Marketing cost analysis The study and evaluation of the relative profitability or costs of different marketing operations in terms of customer, marketing units, commodities, territories, or marketing activities. Typical tools include cost accounting.

Marketing research The systematic gathering, recording, and analyzing of data about problems relating to the marketing of goods and services. Such research may be undertaken by impartial agencies or by business firms or their agents. Marketing research is an inclusive term which includes various subsidiary types:

1. *Market analysis,* of which *product potential* is a type, which is the study of size, location, nature, and characteristics of markets.
2. *Sales analysis* (or *research*), which is the systematic study and comparison of sales (or consumption) data.
3. *Consumer research,* of which *motivation research* is a type that is concerned chiefly with the discovery and analysis of consumer attitudes, reactions, and preferences.

Maximum out-of-pocket cash The highest year-end negative cash balance during project life.

Measures of project worth Economic methods which combine project benefits (savings) and costs in various ways to evaluate the economic value of a project. Examples are life-cycle costs; net benefits or net savings; benefit-to-cost ratio or savings-to-investment ratio; and payback.

Multiple straight-line depreciation method A method of depreciation accounting in which two or more straight-line rates are used. This method permits a predetermined portion of the asset to be written off in a fixed number of years. One common practice is to employ a straight-line rate which will write off 3/4 of the cost in the first half of the anticipated service life, with a second straight-line rate to write off the remaining 1/4 in the remaining half-life.

Net present value (*see* **Present value**)

Net profit Earnings after all operating expenses (cash or accrued noncash) have been deducted from net operating revenues for a given time period.

Net profit, percent of sales (profit margin) The ratio of annual profits to total sales for a representative year of capacity operations. An incomplete measure of profitability, but a useful guidepost for comparing similar products and companies.

Net savings The difference between savings and costs, where both are discounted to present or annual values. The net savings method can be used as a measure of project worth.

Obsolescence (1) The condition of being out of date; a loss of value occasioned by new developments which place the older property at a competitive disadvantage; a factor in depreciation; (2) a decrease in the value of an asset brought about by the development of new and more economical methods, processes, and/or machinery; (3) the loss of usefulness or worth of a product or facility as a result of the appearance of better and/or more economical products, methods, or facilities.

Offsites General facilities outside the battery limits of process units, such as field storage, utilities, and administrative buildings.

On-stream factor The ratio of actual operating days to calendar days per year.

Optimum plant size The plant capacity which represents the best balance between the economics of size and the cost of carrying excess capacity during the initial years of sales.

Overhead A cost or expense inherent in the performing of an operation, i.e., engineering, construction, operating, or manufacturing, which cannot be charged to or identified with a part of the work, product, or asset and therefore must be allocated on some arbitrary base believed to be equitable, or handled as a business expense independent of the volume of production. Plant overhead is also called factory expense.

Payback (payoff) period The period of time at which a machine, facility, or other investment has produced sufficient net revenue to recover its investment costs. Most recent practice is to base payback time on an actual sales projection. Also called payout period. It is simple to calculate and can be used for evaluating many projects. It is not satisfactory for comparing projects with different lives or patterns of cost and earnings.

Present value (present worth) (1) The discounted value of a series of cash flows at an arbitrary point in time. Also the system of comparing proposed investments which involves discounting at a known interest rate (representing a cost of capital or a minimum acceptable rate of return) in order to choose the alternative having the highest present value per unit of investment. This technique eliminates the occasional difficulty with profitability index of multiple solutions, but has the troublesome problem of choosing or calculating a "cost of capital" or minimum rate of return. (2) The time-equivalent value at a specified base time (the present) of past, present, and future cash flows.

Present value (worth) factor (1) The discount factor used to convert future values (benefits and costs) to present values; (2) a mathematical expression also known as the present value of an annuity of 1; (3) one of a set of mathematical formulas used to facilitate calculation of present worth in economic analyses involving compound interest.

Productivity Relative measure of labor efficiency, either good or bad, when compared to an established base or norm as determined from experience.

Productivity changes may be either an increase or decrease in cost. Alternatively, productivity is defined as the reciprocal of **labor factor**.

Profit:

1. *Gross profit*—earnings from an ongoing business after direct costs of goods sold have been deducted from sales revenue for a given period.
2. *Net profit*—earnings or income after subtracting miscellaneous income and expenses (patent royalties, interest, capital gains) and federal income tax from operating profit.
3. *Operating profit*—earnings or income after all expenses (selling, administrative, depreciation) have been deduced from gross profit.

Profitability A measure of the excess of income over expenditure during a given period of time.

Profitability index (PI) The rate of compound interest at which the company's outstanding investment is repaid by proceeds from the project. All proceeds from the project, beyond that required for interest, are credited, by the method of solution, toward repayment of investment by this calculation. Also called discounted cash flow, interest rate of return, investor's method, and internal rate of return. Although frequently requiring more time to calculate than other valid yardsticks, PI reflects in a single number both the dollar and the time values of all money involved in a project. In some very special cases, such as multiple changes of sign in cumulative cash position, false and multiple solutions can be obtained by this technique.

Project life (economic life) Total years of operation for any facility. Sometimes, but not necessarily, equal to depreciable life.

Rate of return on investment The efficiency ratio relating profit or cash flow incomes to investments. Several different measures of this ratio are in common use. (*See also* **Return on average investment, Return on original investment, Profitability index.**)

Replacement A facility proposed to take the place of an existing facility, without increasing its capacity, caused either by obsolescence or physical deterioration.

Required return The minimum return or profit necessary to justify an investment. Often termed interest, expected return or profit, or charge for the use of capital.

Research expense Those continuing expenses required to provide and maintain the facilities to develop new products and improve present products.

Retirement of debt The termination of a debt obligation by appropriate settlement with lender—understood to be in full amount unless partial settlement is specified.

Return on average investment The ratio of annual profits to the average book value of fixed capital, with or without working capital. This method has some advantages over the return-on-original-investment method. Depreciation is always considered; salvage values and recovery of working capital are accounted for. However, the method does not account for the timing of cash flow and yields answers that are considerably higher than those obtained by the return-on-original-investment and profitability index methods. Results may be deceiving when compared, say, against the company's cost of capital.

Return on original investment The ratio of expected average annual aftertax profit (during the earning life) to total investment (working capital included). Similar in usefulness and limitations to payoff period.

Risk The degree of dispersion or variability around the expected or "best" value which is estimated to exist for the economic variable in question; e.g., a quantitative measure of the upper and lower limits which are considered reasonable for the factor being estimated.

Risk exposure The probability that a project's economic outcome is different from what is desired (the target) or what is acceptable.

Sales Orders booked by customers.

Sales forecast A prediction or estimate of sales, in dollars or physical units, for a specified future period under a proposed marketing plan or program and under an assumed set of economic and other forces outside the unit for which the forecast is made. The forecast may be for a specified item of merchandise or for an entire line.

Sales price The revenue received for a unit of product. Gross sales price is the total amount paid. Net sales are gross sales less returns, discounts, freight, and allowances. Plant netbacks are net sales less selling, administrative, and research expenses.

Sales profile The growth or decline of historical or forecast sales volume, by years.

Sales revenue Revenue received as a result of sales, but not necessarily during the same time period.

Salvage value (1) The cost recovered or which could be recovered from a used property when removed, sold, or scrapped; a factor in appraisal of property value and in computing depreciation; (2) the market value of a machine or facility at any point in time. Normally, an estimate of an asset's net market value at the end of its estimated life.

Savings-to-Investment Ratio (SIR) The ratio of present value savings to present value investment costs. The SIR method is used to measure project worth.

Selling expense The total expense involved in marketing the products in question. This normally includes direct selling costs, advertising, and customer service.

Selling price Used interchangeably with sales price.

Sensitivity The relative magnitude of the change in one or more elements of an engineering economy problem that will reverse a decision among alternatives.

Sensitivity analysis A technique for measuring the impact on project outcomes of changing one or more key input values about which there is uncertainty.

Serviceability A measure of the degree to which servicing of an item will be accomplished within a given time under specified conditions.

Servicing The replenishment of consumables needed to keep an item in operating condition, but not including any other preventive maintenance or any corrective maintenance.

Shutdown point The production level at which it becomes less expensive to close the plant and pay remaining fixed expenses out-of-pocket rather than continue operations; that is, the plant cannot meet its variable expense. (*See also* **Breakeven point**.)

Simple interest (1) interest that is not compounded—is not added to the income-producing investment or loan; (2) the interest charges under the condition that interest in any time period is only charged on the principal.

Sinking fund (1) A fund accumulated by periodic deposits and reserved exclusively for a specific purpose, such as retirement of a debt or replacement of a property; (2) a fund created by making periodic deposits (usually equal) at compound interest in order to accumulate a given sum at a given future time for some specific purpose.

Site preparation An act involving grading, landscaping, installation of roads and siding, of an area of ground upon which anything previously located had been cleared so as to make the area free of obstructions, entanglements, or possible collisions with the positioning or placing of anything new or planned.

Startup costs Extra operating costs to bring the plant on stream incurred between the completion of construction and beginning of normal operations. In addition to the difference between actual operating costs during that period and normal costs, it also includes employee training, equipment tests, process adjustments, salaries and travel expense of temporary labor staff, and consultants, report writing, post-startup monitoring, and associated overhead. Additional capital required to correct plant problems may be included. Startup costs are sometimes capitalized.

Straight-line depreciation Method of depreciation whereby the amount to be recovered (written off) is spread uniformly over the estimated (i.e., depreciable) life of the asset in terms of time periods or units of output.

Study period The length of time over which an investment is evaluated, generally set equal to the life of the project or the time horizon of the investor, whichever is shorter.

Sum-of-digits method Also known as **sum-of-the-years-digits** method. A method of computing depreciation in which the amount for any year is based on the ratio: (years of remaining life)/(1 + 2 + 3 ... + n), n being the total anticipated (i.e., depreciable) life.

Sunk cost (1) the unrecovered balance of an investment; it is a cost, already paid, that is not relevant to the decision concerning the future that is being made; capital already invested that for some reason cannot be retrieved; (2) a past cost which has no relevance with respect to future receipts and disbursements of a facility undergoing an economical study. This concept implies that since a past outlay is the same regardless of the alternative selected, it should not influence the choice between alternatives.

Tangibles Things that can be quantitatively measured or valued, such as items of cost and physical assets.

Taxes payable Tax accruals due within a year.

Time value of money (1) The cumulative effect of elapsed time on the money value of an event, based on the earning power of equivalent invested funds (*see also* **Future worth** and **Present worth**); (2) the expected interest rate that capital should or will earn.

Turnover ratio The ratio of annual sales to investment. Inclusion of working capital is preferable, but not always done. Considered by some to be a reasonable basis for a guesstimate of facilities cost, for new products similar to existing products. Ranges around 1.0 for many chemical plants. The product of turnover ratio and profit margin on sales gives a return-on-investment measure.

Uncertainty (or **certainty**) The state of knowledge about the variable inputs to an economic analysis. If the analyst is unsure of input values, there is uncertainty. If the analyst is sure, there is certainty.

Unit cost $/Unit of production. Usually total cost divided by units of production, but a major cost divided by units of production is frequently referred to as a unit cost; for example, the total unit cost is frequently subdivided into the unit cost for labor, chemicals, etc.

Variable costs Raw materials costs, by-product credits, and those processing costs that vary with plant output (such as utilities, catalysts and chemicals, packaging, and labor for batch operations).

Venture life (financial life) The total time span during which expenditures and/or reimbursements related to the venture occur. Venture life may include the research and development, construction, production, and liquidation periods.

Yield The ratio of return or profit over the associated investment, expressed as a percentage or decimal usually on an annual basis. (*See also* **Rate of return on investment.**)

Appendix B

Design and Costing Specifications for Equipment*

AGITATED OPEN TANK
- MATERIAL:
- CAPACITY, volume (gal):
- DIAMETER (ft):
- HEIGHT (ft):
- AGITATOR SPEED (rpm):
- AGITATOR POWER (hp):

AGITATED OPEN TANK, Flotation cell
- MATERIAL:
- CAPACITY, volume per cell (cu ft):
- SINGLE OR DUAL DRIVE:
- DRIVER POWER (hp):

---

* *Source*: U.S. Department of Energy Morgantown Energy Technology Center, METC Fuel Cells Branch, Morgantown, WV. This listing has been adopted by AACE International as a part of its Recommended Practice No. 16R-90, "Conducting Technical and Economic Evaluations in the Process and Utility Industries."

AGITATED PRESSURE TANK
- MATERIAL:
- CAPACITY, volume (gal):
- DIAMETER (ft):
- HEIGHT (ft):
- PRESSURE (psig):
- AGITATOR POWER (hp):

AGITATOR
- MATERIAL:
- CAPACITY (hp):
- SPEED (rpm):
- TYPE IMPELLER:
- TYPE DRIVER:

AIR COMPRESSOR, Centrifugal
- MATERIAL:
- INLET CAPACITY (acfm):
- DISCHARGE PRESSURE (psig):
- INLET TEMPERATURE (°F):
- INLET PRESSURE (psig):
- DRIVER HORSEPOWER (hp):
- TYPE DRIVER:
- STAGES:

AIR DRYER
- MATERIAL:
- INLET CAPACITY (acfm):

BLENDER, Rotary double-cone
- MATERIAL:
- CAPACITY (cu ft):
- SPEED (rpm):
- DRIVER HORSEPOWER (hp):

BLENDER, Rotary drum
- MATERIAL:
- BULK MATERIAL DENSITY (lb/cu ft):
- DRIVER HORSEPOWER (hp):

CENTRIFUGE, ATM suspended basket
- MATERIAL:
- CAPACITY (lb/hr):
- HORSEPOWER (hp):

CENTRIFUGE, Batch automatic
- MATERIAL:

CAPACITY (lb/batch):
CAPACITY (cu ft):
HORSEPOWER (hp):
CENTRIFUGE, Bottom batch
MATERIAL:
CAPACITY (lb/hr):
DIAMETER (in.):
HORSEPOWER (hp):
CENTRIFUGE, Bottom unloading
MATERIAL:
CAPACITY (lb/hr):
DIAMETER (in.):
HORSEPOWER (hp):
CENTRIFUGE, Disk
MATERIAL:
CAPACITY (lb/hr):
DIAMETER (in.):
HORSEPOWER (hp):
CENTRIFUGE, Reciprocating conveyor
MATERIAL:
CAPACITY (lb/hr):
DIAMETER (in.):
CENTRIFUGE, Screen bowl
MATERIAL:
CAPACITY (lb/hr):
DIAMETER (bowl, in.):
LENGTH (bowl, in.):
HORSEPOWER (hp):
CENTRIFUGE, Scroll conveyor
MATERIAL:
CAPACITY (lb/hr):
DIAMETER (in.):
HORSEPOWER (hp):
CENTRIFUGE, Solid bowl
MATERIAL:
CAPACITY (lb/hr):
DIAMETER (bowl, in.):
LENGTH (bowl, in.):
HORSEPOWER (hp):

CENTRIFUGE, Top suspended batch
- MATERIAL:
- CAPACITY (lb/batch):
- DIAMETER (in.):
- HORSEPOWER (hp):

CENTRIFUGE, Top unloading
- MATERIAL:
- CAPACITY (lb/hr):
- DIAMETER (in.):
- HORSEPOWER (hp):

CENTRIFUGE, Tubular
- MATERIAL:
- CAPACITY (lb/hr):
- DIAMETER (in.):
- HORSEPOWER (hp):

CENTRIFUGE, Vibratory
- MATERIAL:
- CAPACITY (tph):
- DIAMETER (in.):
- FEED SIZE (in.):
- HORSEPOWER (hp):

CONVEYOR (Apron, open belt, or closed belt)
- MATERIAL:
- CAPACITY (tons/hr):
- LENGTH (ft):
- WIDTH (in.):
- BULK PRODUCT DENSITY (lbs/cu ft):
- DRIVER HORSEPOWER (hp):

CONVEYOR, Bucket
- MATERIAL:
- CAPACITY (tph):
- LENGTH (ft):
- HORSEPOWER (hp):
- BULK PRODUCT DENSITY (lb/cu ft):
- BUCKET SIZE (e.g., in. width × in. depth):

CONVEYOR, Pneumatic
- MATERIAL:
- CAPACITY (tph):
- LENGTH (ft):
- HORSEPOWER (hp):

BULK PRODUCT DENSITY (lb/cu ft):
LINE SIZE (in.):

CONVEYOR, Roller
MATERIAL:
CAPACITY (tph):
LENGTH (ft):
HORSEPOWER (hp):
DISTANCE BETWEEN CENTERS (in.):
ROLLER SIZE (e.g., in. dia. × in. width):

CONVEYOR, Screw
MATERIAL:
CAPACITY (tph):
LENGTH (ft):
HORSEPOWER (hp):
BULK PRODUCT DENSITY (lb/cu ft):
SCREW DIAMETER (in.):

CONVEYOR, Vibrating
MATERIAL:
CAPACITY (tph):
LENGTH (ft):
PAN WIDTH (in.):

CRANE
MATERIAL:
CAPACITY (tons):
SPAN (ft):
TYPE (bridge or beam):

CRUSHER, Cone
MATERIAL:
CAPACITY (tph):
CONE DIAMETER (in.):
PRODUCT SIZE (in.):
HEAD TYPE (e.g., standard or short):
HORSEPOWER (hp):

CRUSHER, Gyratory
MATERIAL:
CAPACITY (tph):
MANTEL DIAMETER (in.):
PRODUCT SIZE (in.):
TYPE CRUSHING (e.g., primary or secondary):
HORSEPOWER (hp):

CRUSHER, Impact
 CAPACITY (tph):
 FEED OPENING (e.g., 48 in. × 50 in.):
CRUSHER, Jaw
 MATERIAL:
 CAPACITY (tph):
 FEED OPENING SIZE (e.g., 36 in. × 48 in.):
 PRODUCT SIZE (in.):
 HORSEPOWER (hp):
CRUSHER, Reversible hammermill
 MATERIAL:
 CAPACITY (tph):
 FEED OPENING SIZE (e.g., 8 in. × 36 in.):
 HORSEPOWER (hp):
CRUSHER, Roll ring
 MATERIAL:
 CAPACITY (tph):
 FEED OPENING (e.g., 18 in. × 28 in.):
CRUSHER, Rotary
 MATERIAL:
 CAPACITY (tph):
 HORSEPOWER (hp):
CRUSHER, Rotary breaker (Bradford)
 MATERIAL:
 CAPACITY (tph):
 FEED OPENING (in. diameter × in. length):
 PRODUCT SIZE (in.):
 HORSEPOWER (hp):
CRUSHER, Sawtooth
 MATERIAL:
 CAPACITY (tph):
 DRIVER HORSEPOWER (hp):
CRUSHER, Single roll crusher
 MATERIAL:
 CAPACITY (tph):
 ROLL SIZE (e.g., in. diameter × in. length):
CRYSTALLIZER, Batch vacuum
 MATERIAL:
 CAPACITY (tpd):
 CAPACITY (gal):

CRYSTALLIZER, Mechanical
MATERIAL:
CAPACITY (tpd):
LENGTH (ft):
SPEED (rpm):
HORSEPOWER (hp):
CRYSTALLIZER, Oslo
MATERIAL:
CAPACITY (tpd):
DRYER, Atmospheric tray
MATERIAL:
CAPACITY (lb/hr):
AREA OF TOP TRAY (sq ft):
DRYER, DRUM
MATERIAL:
CAPACITY (lb/hr):
PERIPHERAL AREA (sq ft):
SPEED (rpm):
HORSEPOWER (hp):
DRYER, Pan
MATERIAL:
CAPACITY (lb/hr):
AREA (sq ft):
DIAMETER (ft):
DEPTH (in.):
HORSEPOWER (hp):
DRYER, Spray
MATERIAL:
CAPACITY (lb/hr):
EVAPORATIVE CAPACITY (lb water/hr):
DIAMETER (ft):
INLET AIR TEMPERATURE (°F):
DRYER, vacuum tray
MATERIAL:
AREA OF TOP TRAY (sq ft):
DUST COLLECTOR, Centrifugal precipitator
MATERIAL:
CAPACITY (cfm):
GAS TEMPERATURE (°F):

DUST COLLECTOR, Cyclone
MATERIAL:
CAPACITY (cfm):
GAS TEMPERATURE (°F):
DIAMETER (in.):
PRESSURE DROP (in. water gauge):
DUST COLLECTOR, Electrostatic precipitator
MATERIAL:
CAPACITY (cfm):
GAS TEMPERATURE (°F):
PARTICULATE LOADING (ppm):
PARTICULATE COMPOSITION:
VOLTAGE TYPE (e.g., low or high):
DUST COLLECTOR, Multiple cyclone
MATERIAL:
DIAMETER (in.):
DUST COLLECTOR, Washed
MATERIAL:
CAPACITY (cfm):
DIAMETER (in.):
HEIGHT (ft):
EJECTOR
MATERIAL:
CAPACITY (lb/hr):
PUMPING MEDIUM AND PRESSURE (psig and temperature °F):
MEDIUM PUMPED AND PRESSURE (torr):
NUMBER OF STAGES:
ELECTRIC GENERATOR
MATERIAL:
CAPACITY (kVA):
ELEVATOR
CAPACITY (ton):
HEIGHT (ft):
TYPE (freight or passenger):
EVAPORATOR, Agitated falling film
MATERIAL:
CAPACITY (lb/hr):
CAPACITY (gal):
TOTAL HEATING SURFACE AREA (sq ft):
SPEED (rpm):

HORSEPOWER (hp):

EVAPORATOR, Forced circulation
- MATERIAL, SHELL:
- MATERIAL, TUBES:
- CAPACITY (lb/hr):
- CAPACITY (gal):
- TOTAL HEATING SURFACE AREA (sq ft):
- SPEED (rpm):

EVAPORATOR, Long tube film
- MATERIAL, SHELL:
- MATERIAL, TUBES:
- CAPACITY (lb/hr):
- CAPACITY (gal):
- TOTAL HEATING SURFACE AREA (sq ft):
- SPEED (rpm):

EVAPORATOR, Long tube vertical
- MATERIAL, SHELL:
- MATERIAL, TUBE:
- CAPACITY (lb/hr):
- AREA (sq ft):

EVAPORATOR, Standard horizontal tube
- MATERIAL, SHELL:
- MATERIAL, TUBE:
- CAPACITY (lb/hr):
- CAPACITY (gal):
- AREA (sq ft):

EVAPORATOR, Standard vertical tube
- MATERIAL, SHELL:
- MATERIAL, TUBE:
- CAPACITY (lb/hr):
- CAPACITY (gal):
- AREA (sq ft):

EVAPORATOR, Wiped film
- MATERIAL:
- CAPACITY (lb/hr):
- HEAT TRANSFER AREA (sq ft):

FAN, Centrifugal
- MATERIAL:
- CAPACITY (cfm):
- DISCHARGE PRESSURE (psig):

SPEED (rpm):
HORSEPOWER (hp):
TYPE (turbo, propeller, rotary blower, vane-axial, standard industrial):

FEEDER BELT
MATERIAL:
CAPACITY (cu ft/hr):
HORSEPOWER (hp):

FEEDER, Bin-activated
MATERIAL:
CAPACITY (lb/hr):
HORSEPOWER (hp):

FEEDER, Gravimetric
MATERIAL:
CAPACITY (lb/hr):
HORSEPOWER (hp):

FEEDER, Rotary
MATERIAL:
CAPACITY (lb/hr):
DIAMETER (in.):
SPEED (rpm):
HORSEPOWER (hp):

FEEDER, Vibrating
MATERIAL:
CAPACITY (tph):
LENGTH (ft):
WIDTH (in.):
HORSEPOWER (hp):

FILTER, Cartridge
MATERIAL:
CAPACITY (gpm):
PARTICLE RETENTION SIZE (mesh):
OPERATION (manual or automatic):

FILTER, Leaf-dry
MATERIAL:
CAPACITY (lb/batch):
LEAF AREA (sq ft):

FILTER, Pressure leaf-wet
MATERIAL:
CAPACITY (lb/batch):
LEAF AREA (sq ft):

FILTER, Plate and frame
    MATERIAL:
    CAPACITY (lb/batch):
    CAPACITY (frame):
    PLATE SIZE (in. × in.):
FILTER, Rotary disk
    MATERIAL:
    CAPACITY (lb/hr):
    FILTER AREA (sq ft):
    SPEED (rpm):
    HORSEPOWER (hp):
FILTER, Rotary drum
    MATERIAL:
    CAPACITY (lb/hr):
    FILTER AREA (sq ft):
    SPEED (rpm):
    HORSEPOWER (hp):
FILTER, Scroll
    MATERIAL:
    CAPACITY (tph):
    SCREEN DIAMETER (in.):
    FEED SIZE (medium or fine):
FILTER, Sewage
    MATERIAL:
    CAPACITY (lb/hr):
    FILTER AREA (sq ft):
FILTER, Sparkler
    MATERIAL:
    CAPACITY (cu ft):
    FILTER AREA (sq ft):
    DIAMETER (in.):
FLAKER, Drum
    MATERIAL:
    CAPACITY (lb/hr):
    AREA (sq ft):
    SPEED (rpm):
    HORSEPOWER (hp):
FLARE
    MATERIAL:
    CAPACITY (lb/hr):

DIAMETER (in.):
HEIGHT (ft):
TEMPERATURE OF FLARE GAS (°F):
MOLECULAR WEIGHT OF FLARE GAS (lb-moles):
TYPE (guyed, derrick, self-supporting, horizontal):

FURNACE, Heater
MATERIAL:
DUTY (mm btu/hr):
DESIGN PRESSURE (psig):
DESIGN TEMPERATURE (°F):
FUEL FEED RATE (scfm or gpm):
FUEL HEATING VALUE AND TYPE:
TYPE (heater, pyrolysis, reformer, vertical, box):

HEAT EXCHANGER
MATERIAL, SHELL:
MATERIAL, TUBE:
CAPACITY (lb/hr):
HEAT TRANSFER AREA (sq ft):
TUBE LENGTH (ft):
TUBE PRESSURE (psig):
SHELL PRESSURE (psig):
TYPE (floating head, fixed tube sheet, U-tube, cross-bore, graphite tube):

HEAT EXCHANGER, Air cooler
MATERIAL:
BARE TUBE AREA (sq ft):
TUBE LENGTH (ft):
DESIGN PRESSURE (psig):
NUMBER OF BAYS:

HEAT EXCHANGER, Fin tube
MATERIAL:
TUBE LENGTH (ft):
NUMBER OF EXTERNAL FINS:
DESIGN PRESSURE (psig):
NUMBER OF TUBES PER BUNDLE:

HEAT EXCHANGER, Jacketed
MATERIAL:
TUBE LENGTH (ft):
NUMBER OF EXTERNAL FINS:
DESIGN PRESSURE (psig):
NUMBER OF TUBES PER BUNDLE:

HEAT EXCHANGER, Spiral plate
  MATERIAL:
  HEAT TRANSFER AREA (sq ft):
  TUBE PRESSURE (psig):
HEAT EXCHANGER, Suction heater
  MATERIAL:
  HEAT TRANSFER AREA (sq ft):
HEAT EXCHANGER, Tank heater (electric)
  MATERIAL:
  CAPACITY (kW):
HEAT EXCHANGER, Tank heater (steam coil)
  MATERIAL:
  CAPACITY (lb/hr):
  HEAT TRANSFER AREA (sq ft):
  PIPE DIAMETER (ft):
HEAT EXCHANGER, Thermascrew (Reitz)
  MATERIAL:
  HEAT TRANSFER AREA (sq ft):
HEAT EXCHANGER, Two-screw
  MATERIAL:
  HEAT TRANSFER AREA (sq ft):
HEAT EXCHANGER, Waste heat (waste heat boiler)
  MATERIAL:
  CAPACITY (lb/hr):
  HEAT TRANSFER AREA (sq ft):
HEATING UNIT, Dowtherm
  MATERIAL:
  CAPACITY (mm btu/hr):
  CAPACITY (process flow, gpm):
  PRESSURE (psig):
  TEMPERATURE (°F):
HOIST
  LOAD (tons):
  TYPE (single-speed electric, 5-speed electric, plain hand hoist, geared hand hoist):
  WITH OR WITHOUT TROLLEY:
HORIZONTAL TANK, Cylindrical (ASME code)
  MATERIAL:
  CAPACITY (gal):
  DIAMETER (ft):

PRESSURE (psig):
TEMPERATURE (°F):
HORIZONTAL TANK, Multi-wall
MATERIAL:
CAPACITY (gal):
DIAMETER (ft):
LENGTH (ft):
PRESSURE (psig):
TEMPERATURE (°F):
KNEADER
MATERIAL:
CAPACITY (lb/hr):
CAPACITY (gal):
HORSEPOWER (hp):
TYPE (stationary, tilting, vacuum):
LINING
MATERIAL:
LINING AREA (sq ft):
MORTAR TYPE IF BRICK:
TYPE (acid brick, monolithic, other):
TYPE WALL (straight wall tank, small horizontal tank, large horizontal vessel):
MILL
MATERIAL:
CAPACITY (tph):
INSIDE DIAMETER (ft):
INSIDE LENGTH (ft):
DRY OR WET GRINDING:
POWER (hp):
SPEED (rpm):
TYPE (rod, ball, autogenous, attrition, micro-pulverizer, roller):
MIXER
MATERIAL:
CAPACITY (cu ft):
SPEED (rpm):
HORSEPOWER (hp):
TYPE (sigma, fixed propeller, portable propeller, extruder, muller, spiral ribbon, two-roll, pan):
MOTOR
ENCLOSURE:

SPEED (rpm):
HORSEPOWER (hp):
TYPE (open drip-proof, TEFC Class F insulation, explosion-proof, variable speed):

PUMP
MATERIAL:
CAPACITY (gpm):
HEAD (ft):
TEMPERATURE (°F):
LIQUID SPECIFIC GRAVITY:
POWER (hp):
POWER SOURCE (electricity, steam, engine):
TYPE (reciprocating, simplex, duplex, diaphragm, slurry, rotary):

PUMP, Centrifugal
MATERIAL:
CAPACITY (gpm):
HEAD (ft):
TEMPERATURE (°F):
LIQUID SPECIFIC GRAVITY:
POWER (hp):
POWER SOURCE (electricity, steam, engine):
TYPE (single-stage, in-line, vertical, axial flow):

REACTOR
MATERIAL:
CAPACITY (lb/hr):
INSIDE DIAMETER (ft):
TYPE (single-stage, double-stage, fluidized bed):

REBOILER
MATERIAL, SHELL:
MATERIAL, TUBE:
CAPACITY (mm Btu/hr):
HEAT TRANSFER AREA (sq ft):
TUBE LENGTH (ft):
TUBE PRESSURE (psig):
TYPE (kettle, U-tube, thermosiphon):

REFRIGERATION UNIT
MATERIAL:
CAPACITY (tons):
EVAPORATOR TEMPERATURE (°F):
TYPE (mechanical, centrifugal):

ROTARY DRYER
  MATERIAL:
  CAPACITY (lb/hr):
  PERIPHERAL AREA (sq ft):
  SPEED (rpm):
  TYPE (direct, jacketed vacuum, vacuum, indirect)
SCALE
  MATERIAL:
  CAPACITY (lbs):
  BELT WIDTH (in.), "Only belt scale":
  TYPE (beam, semi-frame, full-frame, tank, belt, track, truck):
SEPARATION, Water-only cyclone
  MATERIAL:
  CYCLONE DIAMETER (in.), Individual:
  NUMBER OF CYCLONES:
  LINEAR OR RADIAL CONFIGURATION:
STACK
  MATERIAL:
  HEIGHT (ft):
  DIAMETER (in.):
THICKENER
  MATERIAL (rake):
  DIAMETER (ft):
TOWER
  MATERIAL:
  CAPACITY (lb/hr):
  DIAMETER (ft):
  TRAY SPACING (in.):
  NUMBER OF TRAYS:
  PRESSURE (psig):
TOWER, Cooling
  MATERIAL:
  CAPACITY (gpm):
  COOLING RANGE (°F):
  APPROACH (°F):
  WET BULB TEMPERATURE (°F):
  MAIN HEAD LENGTH(S) (ft):
  SUPPLY AND RETURN LINE LENGTH(S) (ft):
TOWER, Packed
  MATERIAL:

DIAMETER (ft):
PACKING HEIGHT (ft):
PACKING TYPE:
PRESSURE (psig):

TRAY DRYING SYSTEM
MATERIAL:
CAPACITY (lb/hr):
TRAY SURFACE (sq ft):
POWER (hp):
HEATING MEDIUM (steam, air or other):
TYPE (turbo, batch vacuum, atmospheric):

TURBINE, Gas
MATERIAL:
CAPACITY (hp):

TURBINE, Steam
MATERIAL:
CAPACITY (bhp):
SPEED (rpm):
STEAM PRESSURE (psig):
TYPE (condensing or non-condensing):

VACUUM PUMP
MATERIAL:
CAPACITY (inlet cfm):
ULTIMATE PRESSURE (torr):
SPEED (rpm):
POWER (hp):
TYPE (mechanical, water-sealed, mechanical-booster):

VERTICAL TANK, Process
MATERIAL:
CAPACITY (gal):
DIAMETER (ft):
HEIGHT (ft):
PRESSURE (psig):
TEMPERATURE (°F):
TYPE (cylindrical, multi-wall, shell, spheroid, sphere, gas holder):

VERTICAL TANK, Storage
MATERIAL:
VOLUME (gal):
DIAMETER (ft):
HEIGHT (ft):

PRESSURE (psig):
TEMPERATURE (°F):
TYPE (flat bottom/roof, fiberglass, light gauge, cone roof, open top, floating roof, cone bottom bin, lifter):

VIBRATING SCREEN, Rectangular
MATERIAL:
LENGTH (ft):
WIDTH (ft):
ENCLOSURE (yes or no):
POWER (hp):
NUMBER OF DECKS:

VIBRATING SCREEN, Rectangular (Hummer-type)
MATERIAL:
CAPACITY (lb/hr):
SCREEN AREA (sq in):
DEGREE OF SEPARATION (fine or coarse):
NUMBER OF DECKS:

VIBRATING SCREEN, Sifter circular
MATERIAL:
CAPACITY (lb/hr):
SCREEN AREA (sq in.):
SCREEN DIAMETER (in.):
POWER (hp):
NUMBER OF DECKS:

WATER TREATMENT SYSTEM, Boiler
MATERIAL:
CAPACITY (lb/hr):
STEAM PRESSURE (psig):
SATURATED OR SUPERHEATER STEAM:
DEMINERALIZER WATER RATE (gph):
SOFTENING SYSTEM WATER RATE (gph):

Appendix C

Discrete Compound Interest Tables

	Single payment		Uniform annual series	
	Compound interest factor	Present worth factor	Unacost present worth factor	Capital recovery factor
n	$(1+i)^n$	$\frac{1}{(1+i)^n}$	$\frac{(1+i)^n - 1}{i(1+i)^n}$	$\frac{i(1+i)^n}{(1+i)^n - 1}$
	1% Compound Interest Factors			
1	1.010	0.9901	0.990	1.01000
2	1.020	0.9803	1.970	0.50751
3	1.030	0.9706	2.941	0.34002
4	1.041	0.9610	3.902	0.25628
5	1.051	0.9515	4.853	0.20604
6	1.062	0.9420	5.795	0.17255
7	1.072	0.9327	6.728	0.14863
8	1.083	0.9235	7.652	0.13069
9	1.094	0.9143	8.566	0.11674
10	1.105	0.9053	9.471	0.10558
11	1.116	0.8963	10.368	0.09645
12	1.127	0.8874	11.255	0.08885
13	1.138	0.8787	12.134	0.08241
14	1.149	0.8700	13.004	0.07690
15	1.161	0.8613	13.865	0.07212
16	1.173	0.8528	14.718	0.06794
17	1.184	0.8444	15.562	0.06426
18	1.196	0.8360	16.398	0.06098
19	1.208	0.8277	17.226	0.05805
20	1.220	0.8195	18.046	0.05542
21	1.232	0.8114	18.857	0.05303
22	1.245	0.8034	19.660	0.05086
23	1.257	0.7954	20.456	0.04889
24	1.270	0.7876	21.243	0.04707
25	1.282	0.7798	22.023	0.04541
26	1.295	0.7720	22.795	0.04387
27	1.308	0.7644	23.560	0.04245
28	1.321	0.7568	24.316	0.04112
29	1.335	0.7493	25.066	0.03990
30	1.348	0.7419	25.808	0.03875
31	1.361	0.7346	26.542	0.03768
32	1.375	0.7273	27.270	0.03667
33	1.389	0.7201	27.990	0.03573

	Single payment		Uniform annual series	
	Compound interest factor	Present worth factor	Unacost present worth factor	Capital recovery factor
n	$(1+i)^n$	$\frac{1}{(1+i)^n}$	$\frac{(1+i)^n - 1}{i(1+i)^n}$	$\frac{i(1+i)^n}{(1+i)^n - 1}$
	1% Compound Interest Factors			
34	1.403	0.7130	28.703	0.03484
35	1.417	0.7059	29.409	0.03400
40	1.489	0.6717	32.835	0.03046
45	1.565	0.6391	36.095	0.02771
50	1.645	0.6080	39.196	0.02551
	2% Compound Interest Factors			
1	1.020	0.9804	0.980	1.02000
2	1.040	0.9612	1.942	0.51505
3	1.061	0.9423	2.884	0.34675
4	1.082	0.9238	3.808	0.26262
5	1.104	0.9057	4.713	0.21216
6	1.126	0.8880	5.601	0.17853
7	1.149	0.8706	6.472	0.15451
8	1.172	0.8535	7.325	0.13651
9	1.195	0.8368	8.162	0.12252
10	1.219	0.8203	8.983	0.11133
11	1.243	0.8043	9.787	0.10218
12	1.268	0.7885	10.575	0.09456
13	1.294	0.7730	11.348	0.08812
14	1.319	0.7579	12.106	0.08260
15	1.346	0.7430	12.849	0.07783
16	1.373	0.7284	13.578	0.07365
17	1.400	0.7142	14.292	0.06997
18	1.428	0.7002	14.992	0.06670
19	1.457	0.6864	15.678	0.06378
20	1.486	0.6730	16.351	0.06116
21	1.516	0.6598	17.011	0.05878
22	1.546	0.6468	17.658	0.05663
23	1.577	0.6342	18.292	0.05467
24	1.608	0.6217	18.914	0.05287
25	1.641	0.6095	19.523	0.05122

	Single payment		Uniform annual series	
	Compound interest factor	Present worth factor	Unacost present worth factor	Capital recovery factor
n	$(1+i)^n$	$\frac{1}{(1+i)^n}$	$\frac{(1+i)^n - 1}{i(1+i)^n}$	$\frac{i(1+i)^n}{(1+i)^n - 1}$
		2% Compound Interest Factors		
26	1.673	0.5976	20.121	0.04970
27	1.707	0.5859	20.707	0.04829
28	1.741	0.5744	21.281	0.04699
29	1.776	0.5631	21.844	0.04578
30	1.811	0.5521	22.396	0.04465
31	1.848	0.5412	22.938	0.04360
32	1.885	0.5306	23.468	0.04261
33	1.922	0.5202	23.989	0.04169
34	1.961	0.5100	24.499	0.04082
35	2.000	0.5000	24.999	0.04000
40	2.208	0.4529	27.355	0.03656
45	2.438	0.4102	29.490	0.03391
50	2.692	0.3715	31.424	0.03182
		3% Compound Interest Factors		
1	1.030	0.9709	1.03000	0.971
2	1.061	0.9426	0.52261	1.913
3	1.093	0.9151	0.35353	2.892
4	1.126	0.8885	0.26903	3.717
5	1.159	0.8626	0.21835	4.580
6	1.194	0.8375	0.18460	5.417
7	1.230	0.8131	0.16051	6.230
8	1.267	0.7894	0.14246	7.020
9	1.305	0.7664	0.12843	7.786
10	1.344	0.7441	0.11723	8.530
11	1.384	0.7224	0.10808	9.253
12	1.426	0.7014	0.10046	9.954
13	1.469	0.6810	0.09403	10.635
14	1.513	0.6611	0.08853	11.296
15	1.558	0.6419	0.08377	11.938
16	1.605	0.6232	0.07961	12.561
17	1.653	0.6050	0.07595	13.166

	Single payment		Uniform annual series	
	Compound interest factor	Present worth factor	Unacost present worth factor	Capital recovery factor
n	$(1+i)^n$	$\frac{1}{(1+i)^n}$	$\frac{(1+i)^n - 1}{i(1+i)^n}$	$\frac{i(1+i)^n}{(1+i)^n - 1}$
	3% Compound Interest Factors			
18	1.702	0.5874	0.07271	13.754
19	1.754	0.5703	0.06981	14.324
20	1.806	0.5537	0.06722	14.877
21	1.860	0.5375	0.06487	15.415
22	1.961	0.5219	0.06275	15.937
23	1.974	0.5067	0.06081	16.444
24	2.033	0.4919	0.05905	16.936
25	2.094	0.4776	0.05743	17.413
26	2.157	0.4637	0.05594	17.877
27	2.221	0.4502	0.05456	18.327
28	2.288	0.4371	0.05329	18.764
29	2.357	0.4243	0.05211	19.188
30	2.427	0.4120	0.05102	19.600
31	2.500	0.4000	0.05000	20.000
32	2.575	0.3883	0.04905	20.389
33	2.652	0.3770	0.04816	20.766
34	2.732	0.3660	0.04732	21.132
35	2.814	0.3554	0.04654	21.487
40	3.262	0.3066	0.04326	23.115
45	3.782	0.2644	0.04079	24.519
50	4.384	0.2281	0.03887	25.730
	4% Compound Interest Factors			
1	1.040	0.9615	0.962	1.04000
2	1.082	0.9246	1.886	0.53020
3	1.125	0.8890	2.775	0.36035
4	1.170	0.8548	3.630	0.27549
5	1.217	0.8219	4.452	0.22463
6	1.265	0.7903	5.242	0.19076
7	1.316	0.7599	6.002	0.16661
8	1.369	0.7307	6.733	0.14853
9	1.423	0.7026	7.435	0.13449

	Single payment		Uniform annual series	
	Compound interest factor	Present worth factor	Unacost present worth factor	Capital recovery factor
n	$(1+i)^n$	$\frac{1}{(1+i)^n}$	$\frac{(1+i)^n - 1}{i(1+i)^n}$	$\frac{i(1+i)^n}{(1+i)^n - 1}$
		4% Compound Interest Factors		
10	1.480	0.6756	8.111	0.12329
11	1.539	0.6496	8.760	0.11415
12	1.601	0.6246	9.385	0.10655
13	1.665	0.6006	9.986	0.10014
14	1.732	0.5775	10.563	0.09467
15	1.801	0.5553	11.118	0.08994
16	1.873	0.5339	11.652	0.08582
17	1.948	0.5134	12.166	0.08220
18	2.026	0.4936	12.659	0.07899
19	2.107	0.4746	13.134	0.07614
20	2.191	0.4564	13.590	0.07358
21	2.279	0.4388	14.029	0.07128
22	2.370	0.4220	14.451	0.06920
23	2.465	0.4057	14.857	0.06731
24	2.563	0.3901	15.247	0.06559
25	2.666	0.3751	15.622	0.06401
26	2.772	0.3607	15.983	0.06257
27	2.883	0.3468	16.330	0.06124
28	2.999	0.3335	16.663	0.06001
29	3.119	0.3207	16.984	0.05888
30	3.243	0.3083	17.292	0.05783
31	3.373	0.2965	17.588	0.05686
32	3.508	0.2851	17.874	0.05595
33	3.648	0.2741	18.148	0.05510
34	3.794	0.2636	18.411	0.05431
35	3.946	0.2534	18.665	0.05358
40	4.801	0.2083	19.793	0.05052
45	5.841	0.1712	20.720	0.04826
50	7.107	0.1407	21.482	0.04655
		5% Compound Interest Factors		
1	1.050	0.9524	0.952	1.05000

	Single payment		Uniform annual series	
	Compound interest factor	Present worth factor	Unacost present worth factor	Capital recovery factor
n	$(1+i)^n$	$\frac{1}{(1+i)^n}$	$\frac{(1+i)^n - 1}{i(1+i)^n}$	$\frac{i(1+i)^n}{(1+i)^n - 1}$
	5% Compound Interest Factors			
2	1.103	0.9070	1.859	0.53780
3	1.158	0.8638	2.723	0.36721
4	1.216	0.8227	3.546	0.28201
5	1.276	0.7835	4.329	0.23097
6	1.340	0.7462	5.076	0.19702
7	1.407	0.7107	5.786	0.17282
8	1.477	0.6768	6.463	0.15472
9	1.551	0.6446	7.108	0.14069
10	1.629	0.6139	7.722	0.12950
11	1.710	0.5847	8.306	0.12039
12	1.796	0.5568	8.863	0.11283
13	1.886	0.5303	9.394	0.10646
14	1.980	0.5051	9.899	0.10102
15	2.079	0.4810	10.380	0.09634
16	2.183	0.4581	10.838	0.09227
17	2.292	0.4363	11.274	0.08870
18	2.407	0.4155	11.690	0.08555
19	2.527	0.3957	12.085	0.08275
20	2.653	0.3769	12.462	0.08024
21	2.786	0.3589	12.821	0.07800
22	2.925	0.3418	13.163	0.07597
23	3.072	0.3256	13.489	0.07414
24	3.225	0.3101	13.799	0.07247
25	3.386	0.2953	14.094	0.07095
26	3.556	0.2812	14.375	0.06956
27	3.733	0.2678	14.643	0.06829
28	3.920	0.2551	14.898	0.06712
29	4.116	0.2429	15.141	0.06605
30	4.322	0.2314	15.372	0.06505
31	4.538	0.2204	15.593	0.06413
32	4.765	0.2099	15.803	0.06328
33	5.003	0.1999	16.003	0.06249
34	5.253	0.1904	16.193	0.06176

	Single payment		Uniform annual series	
	Compound interest factor	Present worth factor	Unacost present worth factor	Capital recovery factor
n	$(1+i)^n$	$\frac{1}{(1+i)^n}$	$\frac{(1+i)^n - 1}{i(1+i)^n}$	$\frac{i(1+i)^n}{(1+i)^n - 1}$
	5% Compound Interest Factors			
35	5.516	0.1813	16.374	0.06107
40	7.040	0.1420	17.159	0.05828
45	8.985	0.1113	17.774	0.05626
50	11.467	0.0872	18.256	0.05478
	6% Compound Interest Factors			
1	1.060	0.9434	0.943	1.06000
2	1.124	0.8900	1.833	0.54544
3	1.191	0.8396	2.673	0.37411
4	1.262	0.7921	3.465	0.28859
5	1.338	0.7473	4.212	0.23740
6	1.419	0.7050	4.917	0.20336
7	1.504	0.6651	5.582	0.17914
8	1.594	0.6274	6.210	0.16104
9	1.689	0.5919	6.802	0.14702
10	1.791	0.5584	7.360	0.13587
11	1.898	0.5268	7.887	0.12679
12	2.012	0.4970	8.384	0.11928
13	2.133	0.4688	8.853	0.11296
14	2.261	0.4423	9.295	0.10758
15	2.397	0.4173	9.712	0.10296
16	2.540	0.3936	10.106	0.09895
17	2.693	0.3714	10.477	0.09544
18	2.854	0.3503	10.828	0.09236
19	3.026	0.3305	11.158	0.08962
20	3.207	0.3118	11.470	0.08718
21	3.400	0.2942	11.764	0.08500
22	3.604	0.2775	12.042	0.08305
23	3.820	0.2618	12.303	0.08128
24	4.049	0.2470	12.550	0.07968
25	4.292	0.2330	12.783	0.07823
26	4.549	0.2198	13.003	0.07690

	Single payment		Uniform annual series	
	Compound interest factor	Present worth factor	Unacost present worth factor	Capital recovery factor
n	$(1+i)^n$	$\frac{1}{(1+i)^n}$	$\frac{(1+i)^n - 1}{i(1+i)^n}$	$\frac{i(1+i)^n}{(1+i)^n - 1}$
		6% Compound Interest Factors		
27	4.822	0.2074	13.211	0.07570
28	5.112	0.1956	13.406	0.07459
29	5.418	0.1846	13.591	0.07358
30	5.743	0.1741	13.765	0.07265
31	6.088	0.1643	13.929	0.07179
32	6.453	0.1550	14.084	0.07100
33	6.841	0.1462	14.230	0.07027
34	7.251	0.1379	14.368	0.06960
35	7.686	0.1301	14.498	0.06897
40	10.286	0.0972	15.046	0.06646
45	13.765	0.0727	15.456	0.06470
50	18.420	0.0543	15.762	0.06344
		7% Compound Interest Factors		
1	1.070	0.9346	0.935	1.07000
2	1.145	0.8734	1.808	0.55309
3	1.225	0.8163	2.624	0.38105
4	1.311	0.7629	3.387	0.29523
5	1.403	0.7130	4.100	0.24389
6	1.501	0.6663	4.767	0.20980
7	1.606	0.6227	5.389	0.18555
8	1.718	0.5820	5.971	0.16747
9	1.838	0.5439	6.515	0.15349
10	1.967	0.5083	7.024	0.14238
11	2.105	0.4751	7.499	0.13336
12	2.252	0.4440	7.943	0.12590
13	2.410	0.4150	8.358	0.11965
14	2.579	0.3878	8.745	0.11434
15	2.759	0.3624	9.108	0.10979
16	2.952	0.3387	9.447	0.10586
17	3.159	0.3166	9.763	0.10243
18	3.380	0.2959	10.059	0.09941

n	Single payment: Compound interest factor $(1+i)^n$	Single payment: Present worth factor $\frac{1}{(1+i)^n}$	Uniform annual series: Unacost present worth factor $\frac{(1+i)^n - 1}{i(1+i)^n}$	Uniform annual series: Capital recovery factor $\frac{i(1+i)^n}{(1+i)^n - 1}$
	7% Compound Interest Factors			
19	3.617	0.2765	10.336	0.09675
20	3.870	0.2584	10.594	0.09439
21	4.141	0.2415	10.836	0.09229
22	4.430	0.2257	11.061	0.09041
23	4.741	0.2109	11.272	0.08871
24	5.072	0.1971	11.469	0.08719
25	5.427	0.1842	11.654	0.08581
26	5.807	0.1722	11.826	0.08456
27	6.214	0.1609	11.987	0.08343
28	6.649	0.1504	12.137	0.08239
29	7.114	0.1406	12.278	0.08145
30	7.612	0.1314	12.409	0.08059
31	8.145	0.1228	12.532	0.07980
32	8.715	0.1147	12.647	0.07907
33	9.325	0.1072	12.754	0.07841
34	9.978	0.1002	12.854	0.07780
35	10.677	0.0937	12.948	0.07723
40	14.974	0.0668	13.332	0.07501
45	21.002	0.0476	13.606	0.07350
50	29.457	0.0339	13.801	0.07246
	8% Compound Interest Factors			
1	1.080	0.9259	0.926	1.08000
2	1.166	0.8573	1.783	0.56077
3	1.260	0.7938	2.577	0.38803
4	1.360	0.7350	3.312	0.30192
5	1.469	0.6806	3.993	0.25046
6	1.587	0.6302	4.623	0.21632
7	1.714	0.5835	5.206	0.19207
8	1.851	0.5403	5.747	0.17401
9	1.999	0.5002	6.247	0.16008
10	2.159	0.4632	6.710	0.14903

	Single payment		Uniform annual series	
	Compound interest factor	Present worth factor	Unacost present worth factor	Capital recovery factor
n	$(1+i)^n$	$\frac{1}{(1+i)^n}$	$\frac{(1+i)^n - 1}{i(1+i)^n}$	$\frac{i(1+i)^n}{(1+i)^n - 1}$
		8% Compound Interest Factors		
11	2.332	0.4289	7.139	0.14008
12	2.518	0.3971	7.536	0.13270
13	2.720	0.3677	7.904	0.12652
14	2.937	0.3405	8.244	0.12130
15	3.172	0.3152	8.559	0.11683
16	3.426	0.2919	8.851	0.11298
17	3.700	0.2703	9.122	0.10963
18	3.996	0.2502	9.372	0.10670
19	4.316	0.2317	9.604	0.10413
20	4.661	0.2145	9.818	0.10185
21	5.034	0.1987	10.017	0.09983
22	5.437	0.1839	10.201	0.09803
23	5.871	0.1703	10.371	0.09642
24	6.341	0.1577	10.529	0.09498
25	6.848	0.1460	10.675	0.09368
26	7.396	0.1352	10.810	0.09251
27	7.988	0.1252	10.935	0.09145
28	8.627	0.1159	11.051	0.09049
29	9.317	0.1073	11.158	0.08962
30	10.063	0.0994	11.258	0.08883
31	10.868	0.0920	11.350	0.08811
32	11.737	0.0852	11.435	0.08745
33	12.676	0.0789	11.514	0.08685
34	13.690	0.0730	11.587	0.08630
35	14.785	0.0676	11.655	0.08580
40	21.725	0.0460	11.925	0.08386
45	31.920	0.0313	12.108	0.08259
50	46.902	0.0213	12.233	0.08174
		10% Compound Interest Tables		
1	1.100	0.9091	0.909	1.10000
2	1.210	0.8264	1.736	0.57619

	Single payment		Uniform annual series	
	Compound interest factor	Present worth factor	Unacost present worth factor	Capital recovery factor
n	$(1+i)^n$	$\frac{1}{(1+i)^n}$	$\frac{(1+i)^n-1}{i(1+i)^n}$	$\frac{i(1+i)^n}{(1+i)^n-1}$
	10% Compound Interest Factors			
3	1.331	0.7513	2.487	0.40211
4	1.464	0.6830	3.170	0.31547
5	1.611	0.6209	3.791	0.26380
6	1.772	0.5645	4.355	0.22961
7	1.949	0.5132	4.868	0.20541
8	2.144	0.4665	5.335	0.18744
9	2.358	0.4241	5.759	0.17364
10	2.594	0.3855	6.144	0.16275
11	2.853	0.3505	6.495	0.15396
12	3.138	0.3186	6.814	0.14676
13	3.452	0.2897	7.103	0.14078
14	3.797	0.2633	7.367	0.13575
15	4.177	0.2394	7.606	0.13147
16	4.595	0.2176	7.824	0.12782
17	5.054	0.1978	8.022	0.12466
18	5.560	0.1799	8.201	0.12193
19	6.116	0.1635	8.365	0.11955
20	6.727	0.1486	8.514	0.11746
21	7.400	0.1351	8.649	0.11562
22	8.140	0.1228	8.772	0.11401
23	8.954	0.1117	8.883	0.11257
24	9.850	0.1015	8.985	0.11130
25	10.835	0.0923	9.077	0.11017
26	11.918	0.0839	9.161	0.10916
27	13.110	0.0768	9.237	0.10826
28	14.421	0.0693	9.307	0.10745
29	15.863	0.0630	9.370	0.10673
30	17.449	0.0573	9.427	0.10608
31	19.194	0.0521	9.479	0.10550
32	21.114	0.0474	9.526	0.10497
33	23.225	0.0431	9.569	0.10450
34	25.548	0.0391	9.609	0.10407
35	28.102	0.0356	9.644	0.10369

	Single payment		Uniform annual series	
	Compound interest factor	Present worth factor	Unacost present worth factor	Capital recovery factor
n	$(1+i)^n$	$\frac{1}{(1+i)^n}$	$\frac{(1+i)^n-1}{i(1+i)^n}$	$\frac{i(1+i)^n}{(1+i)^n-1}$
		10% Compound Interest Factors		
40	45.259	0.0221	9.779	0.10226
45	72.890	0.0137	9.863	0.10139
50	117.391	0.0085	9.915	0.10086
		12% Compound Interest Factors		
1	1.120	0.8929	0.893	1.12000
2	1.254	0.7972	1.690	0.59170
3	1.405	0.7118	2.402	0.41635
4	1.574	0.6355	3.037	0.32923
5	1.762	0.5674	3.605	0.27741
6	1.974	0.5066	4.111	0.24323
7	2.211	0.4523	4.564	0.21912
8	2.476	0.4039	4.968	0.20130
9	2.773	0.3606	5.328	0.18768
10	3.106	0.3220	5.650	0.17698
11	3.479	0.2875	5.938	0.16842
12	3.896	0.2567	6.194	0.16144
13	4.363	0.2292	6.424	0.15568
14	4.887	0.2046	6.628	0.15087
15	5.474	0.1827	6.811	0.14682
16	6.130	0.1631	6.974	0.14339
17	6.866	0.1456	7.120	0.14046
18	7.690	0.1300	7.250	0.13794
19	8.613	0.1161	7.366	0.13576
20	9.646	0.1037	7.469	0.13388
21	10.804	0.0926	7.562	0.13224
22	12.100	0.0826	7.645	0.13081
23	13.552	0.0738	7.718	0.12956
24	15.179	0.0659	7.784	0.12846
25	17.000	0.0588	7.843	0.12750
26	19.040	0.0525	7.896	0.12665
27	21.325	0.0469	7.943	0.12590
28	23.884	0.0419	7.984	0.12524

n	Single payment: Compound interest factor $(1+i)^n$	Single payment: Present worth factor $\frac{1}{(1+i)^n}$	Uniform annual series: Unacost present worth factor $\frac{(1+i)^n - 1}{i(1+i)^n}$	Uniform annual series: Capital recovery factor $\frac{i(1+i)^n}{(1+i)^n - 1}$
	12% Compound Interest Factors			
29	26.750	0.0374	8.022	0.12466
30	29.960	0.0334	8.055	0.12414
31	33.555	0.0298	8.085	0.12369
32	37.582	0.0266	8.112	0.12328
33	42.091	0.0238	8.135	0.12292
34	47.142	0.0212	8.157	0.12260
35	52.799	0.0189	8.176	0.12232
40	93.051	0.0107	8.244	0.12130
45	163.987	0.0061	8.283	0.12074
50	289.001	0.0035	8.305	0.12042
	15% Compound Interest Factors			
1	1.150	0.8696	0.870	1.15000
2	1.322	0.7561	1.626	0.61512
3	1.521	0.6575	2.283	0.43798
4	1.749	0.5718	2.855	0.35027
5	2.011	0.4972	3.352	0.29832
6	2.313	0.4323	3.784	0.26424
7	2.660	0.3759	4.160	0.24036
8	3.059	0.3269	4.487	0.22285
9	3.518	0.2843	4.772	0.20957
10	4.046	0.2472	5.019	0.19925
11	4.652	0.2149	5.234	0.19107
12	5.350	0.1869	5.421	0.18448
13	6.153	0.1625	5.583	0.17911
14	7.076	0.1413	5.724	0.17469
15	8.137	0.1229	5.847	0.17102
16	9.358	0.1069	5.954	0.16795
17	10.761	0.0929	6.047	0.16537
18	12.375	0.0808	6.128	0.16319
19	14.232	0.0703	6.198	0.16134
20	16.367	0.0611	6.259	0.15976

	Single payment		Uniform annual series	
	Compound interest factor	Present worth factor	Unacost present worth factor	Capital recovery factor
n	$(1+i)^n$	$\frac{1}{(1+i)^n}$	$\frac{(1+i)^n - 1}{i(1+i)^n}$	$\frac{i(1+i)^n}{(1+i)^n - 1}$
	15% Compound Interest Factors			
21	18.821	0.0531	6.312	0.15842
22	21.645	0.0462	6.359	0.15727
23	24.891	0.0402	6.399	0.15628
24	28.625	0.0349	6.434	0.15543
25	32.919	0.0304	6.464	0.15470
26	37.857	0.0264	6.491	0.15407
27	43.535	0.0230	6.514	0.15353
28	50.065	0.0200	6.534	0.15306
29	57.575	0.0174	6.551	0.15265
30	66.212	0.0151	6.566	0.15230
31	76.143	0.0131	6.579	0.15200
32	87.565	0.0114	6.591	0.15173
33	100.700	0.0099	6.600	0.15150
34	115.805	0.0086	6.609	0.15131
35	133.175	0.0075	6.617	0.15113
40	267.862	0.0037	6.642	0.15056
45	538.767	0.0019	6.654	0.15028
50	1083.652	0.0009	6.661	0.15014
	20% Compound Interest Factors			
1	1.200	0.8333	0.833	1.20000
2	1.440	0.6944	1.528	0.65455
3	1.728	0.5787	2.106	0.47473
4	2.074	0.4823	2.589	0.38629
5	2.488	0.4019	2.991	0.33438
6	2.986	0.3349	3.326	0.30071
7	3.583	0.2791	3.605	0.27742
8	4.300	0.2326	3.837	0.26061
9	5.160	0.1938	4.031	0.24808
10	6.192	0.1615	4.192	0.23852
11	7.430	0.1346	4.327	0.23110
12	8.916	0.1122	4.439	0.22526

n	Single payment: Compound interest factor $(1+i)^n$	Single payment: Present worth factor $\frac{1}{(1+i)^n}$	Uniform annual series: Unacost present worth factor $\frac{(1+i)^n - 1}{i(1+i)^n}$	Uniform annual series: Capital recovery factor $\frac{i(1+i)^n}{(1+i)^n - 1}$
	20% Compound Interest Factors			
13	10.699	0.0935	4.533	0.22062
14	12.839	0.0779	4.611	0.21689
15	15.407	0.0649	4.675	0.21388
16	18.488	0.0541	4.730	0.21144
17	22.186	0.0451	4.775	0.20944
18	26.623	0.0376	4.812	0.20781
19	31.948	0.0313	4.844	0.20646
20	38.338	0.0261	4.870	0.20536
21	46.005	0.0217	4.891	0.20444
22	55.206	0.0181	4.909	0.20369
23	66.247	0.0151	4.925	0.20307
24	79.497	0.0126	4.937	0.20255
25	95.396	0.0105	4.948	0.20212
26	114.475	0.0087	4.956	0.20176
27	137.370	0.0073	4.964	0.20147
28	164.845	0.0061	4.970	0.20122
29	197.813	0.0051	4.975	0.20102
30	237.376	0.0042	4.979	0.20085
31	284.851	0.0035	4.982	0.20070
32	341.822	0.0029	4.985	0.20059
33	410.186	0.0024	4.988	0.20049
34	492.223	0.0020	4.990	0.20041
35	590.668	0.0017	4.992	0.20034
40	1469.771	0.0007	4.997	0.20014
45	3657.258	0.0003	4.999	0.20005
50	9100.427	0.0001	4.999	0.20002
	25% Compound Interest Factors			
1	1.250	0.8000	0.800	1.25000
2	1.563	0.6400	1.440	0.69444
3	1.953	0.5120	1.952	0.51230

	Single payment		Uniform annual series	
	Compound interest factor	Present worth factor	Unacost present worth factor	Capital recovery factor
n	$(1+i)^n$	$\frac{1}{(1+i)^n}$	$\frac{(1+i)^n - 1}{i(1+i)^n}$	$\frac{i(1+i)^n}{(1+i)^n - 1}$
		25% Compound Interest Factors		
4	2.441	0.4096	2.362	0.42344
5	3.052	0.3277	2.689	0.37185
6	3.815	0.2621	2.951	0.33882
7	4.768	0.2097	3.166	0.31634
8	5.961	0.1678	3.329	0.30040
9	7.451	0.1342	3.463	0.28876
10	9.313	0.1074	3.571	0.28007
11	11.642	0.0859	3.656	0.27349
12	14.552	0.0687	3.725	0.26845
13	18.190	0.0550	3.780	0.26454
14	22.737	0.0440	3.824	0.26150
15	28.422	0.0352	3.859	0.25912
16	35.527	0.0281	3.887	0.25724
17	44.409	0.0225	3.910	0.25576
18	55.511	0.0180	3.928	0.25459
19	69.389	0.0144	3.942	0.25366
20	86.736	0.0115	3.954	0.25292
21	108.420	0.0092	3.963	0.25233
22	135.525	0.0074	3.971	0.25186
23	169.407	0.0059	3.976	0.25148
24	211.758	0.0047	3.981	0.25119
25	264.698	0.0038	3.985	0.25095
26	330.872	0.0030	3.988	0.25076
27	413.590	0.0024	3.990	0.25061
28	516.988	0.0019	3.992	0.25048
29	646.235	0.0015	3.994	0.25039
30	807.794	0.0012	3.995	0.25031
31	1009.742	0.0010	3.996	0.25025
32	1262.177	0.0008	3.997	0.25020
33	1577.722	0.0006	3.998	0.25016
34	1972.152	0.0005	3.998	0.25013
35	2465.190	0.0004	3.998	0.25010

	Single payment		Uniform annual series	
	Compound interest factor	Present worth factor	Unacost present worth factor	Capital recovery factor
n	$(1+i)^n$	$\frac{1}{(1+i)^n}$	$\frac{(1+i)^n - 1}{i(1+i)^n}$	$\frac{i(1+i)^n}{(1+i)^n - 1}$
	30% Compound Interest Factors			
1	1.300	0.7692	0.769	1.30000
2	1.690	0.5917	1.361	0.73478
3	2.197	0.4552	1.816	0.55063
4	2.856	0.3501	2.166	0.46163
5	3.713	0.2693	2.436	0.41058
6	4.827	0.2072	2.643	0.37839
7	6.275	0.1594	2.802	0.35687
8	8.157	0.1226	2.925	0.34192
9	10.604	0.0943	3.019	0.33124
10	13.785	0.0725	3.091	0.32346
11	17.921	0.0558	3.147	0.31773
12	23.297	0.0429	3.190	0.31345
13	30.287	0.0330	3.223	0.31024
14	39.373	0.0254	3.249	0.30782
15	51.185	0.0195	3.268	0.30598
16	66.540	0.0150	3.283	0.30458
17	86.503	0.0116	3.295	0.30351
18	112.45	0.0089	3.304	0.30269
19	146.18	0.0068	3.310	0.30207
20	190.04	0.0053	3.316	0.30159
21	247.06	0.0040	3.320	0.30122
22	321.17	0.0031	3.323	0.30094
23	417.53	0.0024	3.325	0.30072
24	542.79	0.0018	3.327	0.30055
25	705.62	0.0014	3.329	0.30043
26	917.31	0.0011	3.330	0.30033
27	1192.5	0.0008	3.330	0.30025
28	1550.2	0.0006	3.331	0.30019
29	2015.3	0.0005	3.331	0.30015
30	2619.9	0.0004	3.332	0.30011
31	3405.9	0.0003	3.332	0.30009
32	4427.6	0.0002	3.333	0.30007
33	5755.9	0.0002	3.333	0.30005

	Single payment		Uniform annual series	
	Compound interest factor	Present worth factor	Unacost present worth factor	Capital recovery factor
n	$(1+i)^n$	$\frac{1}{(1+i)^n}$	$\frac{(1+i)^n - 1}{i(1+i)^n}$	$\frac{i(1+i)^n}{(1+i)^n - 1}$
	30% Compound Interest Factors			
34	7482.7	0.0001	3.333	0.30004
35	9727.5	0.0001	3.333	0.30003
	40% Compound Interest Factors			
1	1.400	0.7143	0.714	1.40000
2	1.960	0.5102	1.224	0.81667
3	2.744	0.3644	1.589	0.62936
4	3.842	0.2603	1.849	0.54077
5	5.378	0.1859	2.035	0.49136
6	7.530	0.1328	2.168	0.46126
7	10.541	0.0949	2.263	0.44192
8	14.757	0.0678	2.331	0.42907
9	20.660	0.0484	2.379	0.42034
10	28.925	0.0346	2.414	0.41432
11	40.495	0.0247	2.438	0.41013
12	56.693	0.0176	2.456	0.40718
13	79.370	0.0126	2.469	0.40510
14	111.11	0.0090	2.478	0.40363
15	155.56	0.0064	2.484	0.40259
16	217.79	0.0046	2.489	0.40184
17	304.91	0.0033	2.491	0.40132
18	426.87	0.0023	2.494	0.40094
19	597.62	0.0017	2.496	0.40067
20	836.67	0.0012	2.497	0.40048
21	1171.3	0.0009	2.498	0.40034
22	1639.8	0.0006	2.498	0.40024
23	2295.8	0.0004	2.499	0.40017
24	3214.1	0.0003	2.499	0.40012
25	4499.8	0.0002	2.499	0.40009
	50% Compound Interest Factors			
1	1.500	0.6667	0.667	1.50000

	Single payment		Uniform annual series	
	Compound interest factor	Present worth factor	Unacost present worth factor	Capital recovery factor
n	$(1+i)^n$	$\frac{1}{(1+i)^n}$	$\frac{(1+i)^n - 1}{i(1+i)^n}$	$\frac{i(1+i)^n}{(1+i)^n - 1}$
	50% Compound Interest Factors			
2	2.250	0.4444	1.111	0.90000
3	3.375	0.2963	1.407	0.71053
4	5.063	0.1975	1.605	0.62308
5	7.594	0.1317	1.737	0.57583
6	11.390	0.0878	1.824	0.54812
7	17.085	0.0585	1.883	0.53108
8	25.628	0.0390	1.921	0.52030
9	38.443	0.0260	1.948	0.51335
10	57.665	0.0173	1.965	0.50882
11	86.497	0.0116	1.977	0.50585
12	129.74	0.0077	1.985	0.50388
13	194.61	0.0051	1.990	0.50258
14	291.92	0.0034	1.993	0.50172
15	437.89	0.0023	1.995	0.50114
16	656.84	0.0015	1.997	0.50076
17	985.26	0.0010	1.998	0.50051
18	1477.8	0.0017	1.999	0.50034
19	2216.8	0.0005	1.999	0.50023
20	3325.2	0.0003	1.999	0.50015
	60% Compound Interest Factors			
1	1.600	0.6250	0.625	1.60000
2	2.560	0.3906	1.016	0.98462
3	4.096	0.2441	1.260	0.79380
4	6.554	0.1526	1.412	0.70804
5	10.485	0.0954	1.508	0.66325
6	16.777	0.0596	1.567	0.63803
7	26.843	0.0373	1.604	0.62322
8	42.949	0.0239	1.628	0.61430
9	68.719	0.0146	1.642	0.60886
10	109.95	0.0091	1.652	0.60551
11	175.92	0.0057	1.657	0.60343

	Single payment		Uniform annual series	
	Compound interest factor	Present worth factor	Unacost present worth factor	Capital recovery factor
n	$(1+i)^n$	$\frac{1}{(1+i)^n}$	$\frac{(1+i)^n - 1}{i(1+i)^n}$	$\frac{i(1+i)^n}{(1+i)^n - 1}$
	60% Compound Interest Factors			
12	281.47	0.0036	1.661	0.60214
13	450.35	0.0022	1.663	0.60134
14	720.57	0.0014	1.664	0.60083
15	1152.9	0.0009	1.665	0.60052
16	1844.6	0.0005	1.666	0.60033
17	2951.4	0.0003	1.666	0.60020
18	4722.3	0.0002	1.666	0.60013
19	7555.7	0.0001	1.666	0.60008
20	12089.0	0.0001	1.667	0.60005
	70% Compound Interest Factors			
1	1.700	0.5882	0.588	1.7000
2	2.890	0.3460	0.934	1.0703
3	4.913	0.2035	1.138	0.87889
4	8.352	0.1197	1.258	0.79521
5	14.198	0.0704	1.328	0.75304
6	24.137	0.0414	1.369	0.73025
7	41.033	0.0244	1.394	0.71749
8	69.757	0.0143	1.408	0.71018
9	118.58	0.0084	1.417	0.70595
10	201.59	0.0050	1.421	0.70349
11	342.71	0.0029	1.424	0.70205
12	582.62	0.0017	1.426	0.70120
13	990.45	0.0010	1.427	0.70071
14	1683.7	0.0006	1.428	0.70042
15	2862.4	0.0004	1.428	0.70024
16	4866.0	0.0002	1.428	0.70014
17	8272.3	0.0001	1.428	0.70008
18	14063.0	0.0001	1.428	0.70005
19	23907.0	–	1.429	0.70003
20	40642.0	–	1.429	0.70002

	Single payment		Uniform annual series	
	Compound interest factor	Present worth factor	Unacost present worth factor	Capital recovery factor
n	$(1+i)^n$	$\frac{1}{(1+i)^n}$	$\frac{(1+i)^n - 1}{i(1+i)^n}$	$\frac{i(1+i)^n}{(1+i)^n - 1}$
		80% Compound Interest Factors		
1	1.800	0.5556	0.556	1.8000
2	3.240	0.3086	0.864	1.1571
3	5.831	0.1715	1.036	0.96556
4	10.497	0.0953	1.313	0.88423
5	18.895	0.0529	1.184	0.84470
6	34.012	0.0294	1.213	0.82423
7	61.221	0.0163	1.230	0.81328
8	110.19	0.0091	1.239	0.80733
9	198.35	0.0050	1.244	0.80405
10	357.04	0.0028	1.247	0.80225
11	642.68	0.0016	1.248	0.80125
12	1156.8	0.0009	1.249	0.80069
13	2082.2	0.0004	1.249	0.80038
14	3748.1	0.0003	1.250	0.80021
15	6746.5	0.0001	1.250	0.80012
		90% Compound Interest Factors		
1	1.900	0.5263	0.526	1.9000
2	3.610	0.2770	0.803	1.2448
3	6.859	0.1458	0.949	1.0536
4	13.032	0.0767	1.026	0.97480
5	24.760	0.0404	1.066	0.93788
6	47.045	0.0213	1.087	0.91955
7	89.386	0.0112	1.099	0.91018
8	169.83	0.0059	1.105	0.90533
9	322.68	0.0031	1.108	0.90280
10	613.10	0.0016	1.109	0.90147
		100% Compound Interest Factors		
1	2.000	0.5000	0.500	2.0000
2	4.000	0.2500	0.750	1.3333

	Single payment		Uniform annual series	
	Compound interest factor	Present worth factor	Unacost present worth factor	Capital recovery factor
n	$(1+i)^n$	$\frac{1}{(1+i)^n}$	$\frac{(1+i)^n - 1}{i(1+i)^n}$	$\frac{i(1+i)^n}{(1+i)^n - 1}$
	100% Compound Interest Factors			
3	8.000	0.1250	0.875	1.1428
4	16.000	0.0625	0.938	1.0666
5	32.000	0.0313	0.969	1.0322
6	64.00	0.0156	0.984	1.0158
7	128.00	0.0078	0.992	1.0078
8	256.00	0.0039	0.996	1.0039
9	512.00	0.0020	0.998	1.0019
10	1024.00	0.0010	0.999	1.0009

Appendix D

Bibliography*

The literature for cost engineering and estimating is generally quite perishable; costs simply do not remain constant and cost data rapidly becomes obsolete.

This bibliography attempts to get by that problem by including only reference works which are essentially timeless in nature or which are frequently updated.

Articles which have appeared in the periodical literature in the past are intentionally omitted because of their perishable nature. However, the major periodical journals, magazines, and conference proceedings are shown.

The bibliography is divided into three groupings:

---

* Many of these publications are available to members of AACE International (formerly the American Association of Cost Engineers) at substantial discounts. For information concerning these books, as well as membership in AACE, contact the Association at PO Box 1557, Morgantown, WV 26507-1557, USA. Phone (800) 858-COST or (304) 296-8444.

Cost Data Books,
Periodical Literature, and
Reference Books.

No effort has been made to identify each publication as to the specific subjects it covers. In most cases, the publications are very extensive and may deal with a great many cost engineering topics. This is particularly true of the listed periodicals. In the great majority of cases, however, the content of the publication is fairly clearly defined by its title.

Some of the references are published in a language other than English. The language of any such publication is noted in brackets. In some cases, the publication may be partially in English and partially in another language. This also is noted.

In addition to this bibliography, the reader is referred to the reference list at the end of Chapter 8 which covers international cost information sources.

COST DATA BOOKS

Building Construction Cost Data, R.S. Means Co., Kingston, MA, published annually.

Commercial Square Foot Building Costs, Saylor Publications, Inc., Chatsworth, CA, published annually.

Current Construction Costs, Saylor Publications, Inc., Chatsworth, CA, published annually.

Dutch Association of Cost Engineers Prijzenboekje (*Price Book*) [Dutch], Dutch Association of Cost Engineers, Leidschendam, Netherlands, updated every few years.

General Construction Estimating Standards, Richardson Engineering Services, Inc., Mesa, AZ, published annually.

Means Assemblies Cost Data, R.S. Means Co., Kingston, MA, published annually.

Means Concrete and Masonry Cost Data, R.S. Means Co., Kingston, MA, published annually.

Means Electrical Cost Data, R.S. Means Co., Kingston, MA, published annually.

Means Facilities Construction Cost Data, R.S. Means Co., Kingston, MA, published annually.

Means Facilities Maintenance and Repair Cost Data, R.S. Means Co., Kingston, MA, published annually.

Means Heavy Construction Cost Data, R.S. Means Co., Kingston, MA, published annually.

Means Interior Cost Data, R.S. Means Co., Kingston, MA, published annually.

Means Labor Rates for the Construction Industry, R.S. Means Co., Kingston, MA, published annually.

Means Light Commercial Cost Data, R.S. Means Co., Kingston, MA, published annually.

Means Mechanical Cost Data, R.S. Means Co., Kingston, MA, published annually.

Means Site Work and Landscape Cost Data, R.S. Means Co., Kingston, MA, published annually.

Means Square Foot Costs, R.S. Means Co., Kingston, MA, published annually.

National Building Cost Manual, Craftsman Book Co., Carlsbad, CA, published annually.

National Construction Estimator, Craftsman Book Co., Carlsbad, CA, published annually.

National Electrical Estimator, Craftsman Book Co., Carlsbad, CA, published annually.

National Plumbing and HVAC Estimator, Craftsman Book Co., Carlsbad, CA, published annually.

National Repair and Remodeling Estimator, Craftsman Book Co., Carlsbad, CA, published annually.

National Painting Cost Estimator, Craftsman Book Co., Carlsbad, CA, published annually.

Ottaviano, V.B., *National Mechanical Estimator*, Fairmont Press, Lilburn, GA, updated regularly.

Process Plant Construction Estimating Standards, Richardson Engineering Services, Inc., Mesa, AZ, published annually.

Remodeling/Repair Construction Costs, Saylor Publications, Inc., Chatsworth, CA, published annually.

Shenzhen Property and Building Price Yearbook [English and Chinese], Shenzhen Construction Quotation Price Management Department, Shenzhen City, Guangdong Province, People's Republic of China, published annually.

PERIODICAL LITERATURE

AACE Transactions, AACE International, Morgantown, WV, published annually.

ÁR Monitor (**Cost Monitor**) [Hungarian], Gépipari Tudományos Egyesület/ Mûszaki Költségtervezõ Klub, Budapest, a periodic journal.

The Building Economist, Australian Institute of Quantity Surveyors, National Surveyors House, 27-29 Napier Close, Deakin, ACT 2600, Australia, a quarterly journal.

Construction Cost Index, Amax Australia Ltd., 200 St. George's Terrace, Perth, WA 6000, Australia, a monthly magazine.

Cost Engineering, AACE International, Morgantown, WV, a monthly journal of cost estimating, cost control, and project management.

Engineering Costs and Production Economics (quarterly journal), Elsevier Scientific Publishing Co., Amsterdam.

Engineering News-Record, "Cost Roundup," McGraw-Hill, New York, published quarterly.

The Estimator, American Society of Professional Estimators, Wheaton, MD, a bi-monthly magazine.

Hanscomb/Means International Construction Cost Intelligence Report Hanscomb Associates, Atlanta, GA, a bimonthly newsletter.

Ingegneria Economica [English and Italian], Associazione Italiana di Ingegneria Economica, Milan, Italy, a quarterly journal.

Ingenieria de Costos [Spanish], Sociedad Mexicana de Ingeniería Económica, Financiera y de Costos, Mexico City, a quarterly journal.

International Journal of Project Management, Butterworth-Heinemann Ltd., Letchworth, Herts., United Kingdom, a quarterly journal.

La Cible, Le Journal du Management de Projet [French], Association Française des Ingénieurs et Techniciens d'Estimation, de Planification et de Projets, Paris, a monthly journal.

National Estimator, Society of Cost Estimating and Analysis, Alexandria, VA, a quarterly magazine.

PM NETwork, Project Management Institute, Drexel Hill, PA, a monthly magazine.

Producer Price Indexes, U.S. Department of Labor, Bureau of Labor Statistics, Washington, DC, issued monthly.

Project Management Institute Seminar/Symposium Proceedings, Project Management Institute, Drexel Hill, PA, published annually.

Project Management Journal, Project Management Institute, Drexel Hill, PA, a quarterly journal.

Project Pro, Project Publishing, Pretoria, South Africa, a monthly journal.

Prosjektledelse [English and Norwegian], Norsk Forening for Prosjektledelse, Oslo, Norway, a quarterly journal.

Transactions of the International Cost Engineering Congress, published bi-annually by various member societies of the International Cost Engineering Council.

REFERENCE BOOKS

AACE Cost Engineers' Notebook, AACE International, Morgantown, WV, updated and expanded periodically.

AACE Recommended Practices and Standards, AACE International, Morgantown, WV, updated and expanded periodically.

Antill, J.M. and R.W. Woodhead, *Critical Path Methods in Construction Practice*, 4th ed., Wiley-Interscience, New York, 1990.

Ármenedzser Kézikönyv (*Cost Manager Handbook*) [Hungarian], Perfekt, Budapest, 1993.

Bent, J.A. and K.K. Humphreys, *Applied Cost and Schedule Control*, 2nd ed., Marcel Dekker, Inc., New York, (in press, 1996).

Black, J.H., *Cost Engineering Management Techniques*, Marcel Dekker, Inc., New York, 1984.

Blanchard, F.L., *Engineering Project Management*, Marcel Dekker, Inc., New York, 1990.

Bledsoe, J.D., *Successful Estimating Methods*, R.S. Means Co., Kingston, MA, 1992.

Building Estimator's Reference Book, Frank R. Walker Co., Lisle, IL, updated every few years.

Bunea, S.P., *Means Structural Steel Estimating*, R.S. Means Co., Kingston, MA, 1986.

Clough, R.H. and G.A. Sears, *Construction Project Management*, 3rd ed., John Wiley & Sons, New York, 1991.

Clark, F.D. and Lorenzoni, A.B., *Applied Cost Engineering*, 2nd ed., Marcel Dekker, Inc., New York, 1985.

Cook, P.J., *Bidding for Contractors*, R.S. Means Co., Kingston, MA, 1991.

Cook, P.J., *Estimating for Contractors*, R.S. Means Co., Kingston, MA, 1991.

Cook, P.J., *Quantity Takeoff for Contractors*, R.S. Means Co., Kingston, MA, 1991.

Creese, R.C. et al., *Estimating and Costing for the Metal Manufacturing Industries*, Marcel Dekker, Inc., New York, 1992.

Davis Langdon & Seah International (ed.), *Asia Pacific Construction Costs Handbook*, E&FN Spon, London, 1993. Also distributed by R.S. Means Co., Kingston, MA.

Davis Langdon & Everest (ed.), *European Construction Costs Handbook*, E&FN Spon, London, 1992. Also distributed by R.S. Means Co., Kingston, MA.

Diamant, L. and C.R. Tumblin, *Construction Cost Estimates*, 2nd ed., Wiley-Interscience, New York, 1990.

Drigani, F., *Computerized Project Control*, Marcel Dekker, Inc., New York, 1989.

Estimating Check List for Offshore Projects, Association of Cost Engineers, Sandbach, Cheshire, United Kingdom, 1982.

Európai Dollárpiac-Ipari Árarányok (*European USD Market-Industrial Price Ratios)* (Hungarian), Press-Ker, Budapest, 1991.

Gareis, R.(ed.), *Handbook of Management by Projects*, MANZsche Verlags- und Universitatsbuchhandlung, Vienna, 1990.

Gutman, N., *How to Keep Product Costs in Line*, Marcel Dekker, Inc., New York, 1985.

Hackney, J.W., *Control and Management of Capital Projects*, 2nd ed., K.K. Humphreys (ed.), McGraw-Hill, New York, 1992.

Hedden, R.P., *Cost Engineering in Printed Circuit Board Manufacturing*, Marcel Dekker, Inc., New York, 1987.

Horsley, F.W. and B.J. Cox, *Square Foot Estimating*, R.S. Means Co., Kingston, MA, 1983.

Humphreys, K.K. (ed.), *Jelen's Cost and Optimization Engineering*, 3rd ed., McGraw-Hill, New York, 1991.

Humphreys, K.K. and L.M. English (eds.), *Project and Cost Engineers' Handbook*, 3rd ed., Marcel Dekker, Inc., New York, 1993.

Humphreys, K.K. and J. Gladstone (eds.), *Gladstone's Mechanical Estimating Guidebook*, 6th ed., McGraw-Hill, New York, 1995.

Industrial Engineering Terminology, revised ed., Industrial Engineering and Management Press, Norcross, GA 1991.

Kerzner, H., *Project Management: A Systems Approach to Planning, Scheduling and Control*, Van Nostrand Reinhold, New York, 1989.

Kezsbom, D.S., D.L. Schilling, and K.A. Edward, *Dynamic Project Management. A Practical Guide for Managers and Engineers*, John Wiley & Sons, New York, 1989.

Kharbanda, O.P., *Process Plant and Equipment Cost Estimation*, Craftsman Book Co., Carlsbad, CA, 1979.

Kharbanda, O.P. and E.A. Stallworthy, *How to Learn from Project Disasters*, Gower Publishing Co., Hampshire, UK, 1983.

Kharbanda, O.P., E.A. Stallworthy, and L.F. Williams, *Project Cost Control in Action*, 2nd ed., Gower Press, Brookfield, VT, 1987.

Kimmons, R.C., *Project Management Basics*, Marcel Dekker, Inc., New York, 1990.

Kimmons, R.C. and J.H. Loweree, *Project Management*, Marcel Dekker, New York, 1989.

Lang, H.J., *Cost Analysis for Capital Investment Decisions*, Marcel Dekker, Inc., New York, 1989.

Malstrom, E.M., *Manufacturing Cost Engineering Handbook*, Marcel Dekker, Inc., New York, 1984.

Manual for Construction Cost Estimating, Frank R. Walker Co., Chicago, IL, 1981.

Marshall, H.E., *Techniques for Treating Uncertainty and Risk in the Economic Evaluation of Building Investments*, Special Publication 757, U.S. Department of Commerce, National Institute of Standards and Technology, Gaithersburg, MD, 1988.

MasterFormat, Master List of Titles and Numbers for the Construction Industry, Construction Specifications Institute, Alexandria, VA, 1988.

Means Electrical Estimating, R.S. Means Co., Kingston, MA, 1986.

Means Estimating Handbook, R.S. Means Co., Kingston, MA, 1990.

Means Facilities Maintenance Standards, R.S. Means Co., Kingston, MA, 1988.

Means Heavy Construction Handbook, R.S. Means Co., Kingston, MA, 1993.

Means Productivity Standards for Construction, 3rd ed., R.S. Means Co., Kingston, MA, 1994.

Michaels, J.V. and W.P. Wood, *Design to Cost*, John Wiley & Sons, New York, 1989.

Moylan, J.J. and M. Mossman (eds.), *Means Mechanical Estimating: Standards and Procedures*, 2nd ed., R.S. Means, Kingston, MA, 1992.

Navarette, P.F., *Planning, Estimating, and Control of Chemical Plant Construction Projects*, Marcel Dekker, Inc., New York, 1995.

O'Brien, J.J., *Preconstruction Estimating: Budget through Bid*, McGraw-Hill, New York, 1994.

Page, J.S., *Estimators Man-Hour Manual on Heating, Air Conditioning, Ventilating, and Plumbing*, 2nd ed., Gulf Publishing Co., Houston, TX, 1978.

Park, W.R. and W.B. Chapin, Jr., *Construction Bidding: Strategic Pricing for Profit* 2nd ed., John Wiley & Sons, New York, 1992.

Patrascu, A., *Construction Cost Engineering*, Craftsman Book Co., Carlsbad, CA, 1978.

Patrascu, A., *Construction Cost Engineering Handbook*, Marcel Dekker, Inc., New York, 1988.

Richter, I.E., *International Construction Claims: Avoiding and Resolving Disputes*, McGraw-Hill Book Co., New York, 1983.

Riggs, H.E., *Financial and Cost Analysis for Engineering and Technology Management*, John Wiley & Sons, New York, 1994.

Ruegg, R.T. and H.E. Marshall, *Building Economics: Theory and Practice*, Chapman and Hall, New York, 1990.

Samid, G., *Computer-Organized Cost Engineering*, Marcel Dekker, Inc., New York, 1990.

SCCS, Standard Cost Coding System, Norwegian Petroleum Directorate, Statoil, Saga Petroleum, and Norsk Hydro, Sandvika, Norway, May 1992.

Selg, R.A. (ed.), *Hazardous Waste Cost Control*, Marcel Dekker, Inc., New York, 1993.

Shillito, M.L. and D.J. De Marle, *Value—Its Measurement, Design and Management*, John Wiley & Sons, New York, 1992.

Sims, E.R., Jr., *Precision Manufacturing Costing*, Marcel Dekker, Inc., New York, 1995

Sinclair, D.N., *Estimating for Abnormal Conditions*, Industrial Press, Inc., New York, 1989.

Skills and Knowledge of Cost Engineering, 3rd ed., AACE International, Morgantown, WV, 1992.

Stallworthy, E.A., *International Construction and the Role of Project Management*, Gower Press, Brookfield, VT, 1985.

Stukhart, G., *Construction Materials Management*, Marcel Dekker, Inc., New York, 1995.

Thorne, H.C. and Piekarski, J.A., *Techniques for Capital Expenditure Analysis*, Marcel Dekker, Inc., New York, 1995.

Ward, S.A., *Cost Engineering for Effective Project Control*, John Wiley & Sons, New York, 1992.

Westney, R.E., *Computerized Management of Multiple Small Projects*, Marcel Dekker, Inc., New York, 1992.

Appendix E

Answers to Selected Problems

Chapter 2

2.1	Cost = $168,700
2.3	a) $14,880,000
	b) $22,752,000
2.17	Cost ≈ $25,500,000
2.20	Cost ≈ $1,200,000

Chapter 8

8.4	Central Africa: $20,000,000
	New Zealand: $13,000,000
8.6	United States: 16,800 workhours

Colombia: 51,200 workhours
Venezuela: 33,600 workhours

Chapter 10

10.2 Shutdown point: 12.5% of capacity
Breakeven point: 47.3% of capacity
10.4 $5,500,000 per year
10.5 Year 1: $12,000,000
Year 2: $9,600,000
Year 3: $7,680,000
Year 4: $6,144,000
Year 5: $4,915,200

Chapter 12

12.4 DCF = 35.3%
Payback time = 2.22 years
ROI = 25%
Venture worth = $1,729

Chapter 13

13.1 $3,137

Index

AACE, 7, 27, 125, 134, 253, 277, 319
AACE Bulletin, 174
Accelerated cost recovery system (*see* Depreciation)
ACE, 118
ACRS (*see* Depreciation)
Administrative costs, 36
American Association of Cost Engineers (*see* AACE)
American National Standards Institute, 218
Amortization, 202
ANSI, 218
As-built drawings, 97
Association of Cost Engineers (*see* ACE)

Barry, J. E., 125, 134
Barry, W. R., 61
Battery limits, 37
Bibliography, 319–329
Black, J. H., 163–164, 184
Breakeven point, 156–158
Bulk accounts, 3–4, 50, 55
Burden, 181
Business Roundtable, 174
By-products, 160–161

Capacity, reduced, 152–153, 156–158
Capital, cost of, 218–219
Capital cost estimating, 3–138
  importance of, 109–114
  organizing, 47–60
Capital cost summary, 32–35, 54
Capital recovery factor, 224
  tables of, 296–317

Cash flow, 213–215, 219–226
Chemical Engineering plant cost index, 18–19
Chemical plant location factors, international, 121–124
Code of accounts, 48–53, 63–74
Commerce, U.S. Department of, 242
Compound interest factor, 224
  tables of, 296–317
Concrete, 75–76
Construction Specifications Institute, 49, 61–93
Construction Users Anti-Inflation Roundtable, 174
Contingency, 36, 56, 112–113, 150, 182
Control systems, 98
Conveying systems, 86
Cost capacity exponents, 10–18
Cost capacity factors, 8–18
Cost engineering literature, 120, 126–130, 135–138, 319–329
Cost indexes, 9, 18–20, 125–129
  international, 125–129
  limitations of, 19
Cost reports, 245–248
CPM, 111
Critical path method, 111

Databases, international costs, 130
Data sources, international, 120–121
DCF (*see* Discounted cash flow)
Debugging costs, 207–209
Depletion, 203
Depreciation, 37, 183–202, 217–218
  accelerated cost recovery (ACRS), 184, 195–199
  double-declining balance, 193–194
  guidelines, IRS, 184–192
  modified accelerated cost recovery (MACRS), 184–192, 199–202
  straight-line, 193
[Depreciation]
  sum-of-years-digits, 194–195
Direct costs, 142, 149
Discounted cash flow, 215, 220, 222–226
Distribution costs, 203–204
Doors, 79–80

EEC, 8
Electrical summary, 144
Electrical systems, 89–91
Electricity costs, 162–164
Electric power source, 97
Energy, U.S. Department of, 277
Engineering News-Record indexes, 18–19
Equipment
  cost of, 24–31
  installation costs, 26–31, 81–85
  list, detailed, 22–23
  specifications for, 277–294
  summary, 38–39
Estimate
  budget authorization, 7–8, 42–43, 59, 103–104
  conceptual, 61–93
  definitive, 7–8, 43, 59, 104
  detailed, 8, 43–44, 59, 104–105
  factored, 8, 20–41, 58, 102–103
  grass roots, 37
  green field, 37
  order-of-magnitude, 7–20, 48, 57–58, 102
  preliminary, 7–8, 42–43, 59, 103–104
  project control, 8, 43
  ratio, 8–20, 48
  rules of thumb, 74–91
  study, 7–8, 20–41, 58
Estimating forms, operating cost, 152, 154–155
European Chemical News, 159
European Economic Community (*see* EEC)

Factors
- chemical plants, international, 121–124
- cost capacity, 8–18
- distributive percentage, 26–29
- labor, 30
- labor productivity, international, 121–124
- Lang, 25, 28

Factory supplies, 180–181
Field indirects, 53, 56
Financial analysis, 213–244
Finishes, 80–81
Flowsheet, 20–21, 151
Fringe benefits, 181
Fuel costs, 163–164
Furnishings, 85

GATT, 115
General Agreement on Tariffs and Trade (*see* GATT)
General and administrative expense, 150
General requirements, 74–75
General works expense, 150, 183
Grass roots estimate, 37
Green field estimate, 37

Home office costs, 56
Humphreys, K. K., 115, 183
Hurdle rate, 218

ICEC, 125, 131–134
- address, 133
- member societies, 131–133

Incremental present worth (*see* Venture worth analysis)
Indirect costs, 31, 53, 56, 150
Indirects, field, 31, 53, 56
Inflation, impact of, 219
Insurance, 203
Interest during construction, 37
Interest rate of return (*see* Discounted cash flow)
Interior, U.S. Department of, 21
Internal Revenue Service, 184–193, 196, 199, 217–218
International cost data, 115–138
International Cost Engineering Council (*see* ICEC)
International Project Management Association (*see* INTERNET)
INTERNET, 125, 131–133
- address, 133
- member societies, 131–133

IPMA (*see* INTERNET)
Iron Age, 159
IRS (*see* Internal Revenue Service)

Katell, S., 18

Labor
- costs, 101–106, 164–179
- direct, 142–143
- planning, 105–106
- productivity, 174–178
- staffing table, 146–147, 166–173

Labor productivity factors, international, 121–124
Lang, H. J., 25
Lang factors, 25
LCC (see Life-cycle costing)
Life-cycle costing, 229–235
Location factors, international, 119–122

MACRS (*see* Depreciation)
Maintenance costs, 178–180
Manufacturing cost estimates, 141–205
MARR (*see* Minimum acceptable rate of return)
Marshall and Stevens index, 18
Marshall and Swift index, 18–19
Marshall, H. E., 229
Masonry, 76–77

MasterFormat, 49, 61–93
McMillan, B. G., 115
Mechanical systems, 86–89
Metals, 77–78
Mines, U.S. Bureau of, 13–18, 21
Minimum acceptable rate of return, 218
Modified accelerated cost recovery (*see* Depreciation)
Moisture protection, 79
Motor control centers, 97

NAFTA, 8
National Institute of Standards and Technology, 229, 242
Natural gas costs, 163
Nelson-Farrar refinery construction index, 18–19
Net benefits, 235–236
Net present value (*see* Venture worth analysis)
Net savings, 235–236
North American Free Trade Agreement (*see* NAFTA)

Offsites, 4, 48
Oil Paint and Drug Reporter, 159
Operating cost estimates, 141–205
Operating supplies, 180–181
Overhead, 36, 180–182
  laboratory, 181, 182
  payroll, 181
Overtime, 165, 174–178
  scheduled, 174–178

Packaging costs, 203–204
PAICE Associates, Inc., 213
Patrascu, A., 118
Payback time, 219–221, 238
Payout time (*see* Payback time)
Peters, M. S., 12–13, 18
Petersen, S. R., 229
Piekarski, J. A., 213
Pipe-rack loading, 98
Piping and instrument diagrams, 42–43
Plant expansion, 95–100
Plant safety considerations, 96–97
Plastics, 78
Present worth, incremental (*see* Venture worth analysis)
Present worth factor, 224
  tables of, 296–317
  unacost (uniform annual cost), 224
Probability curve, 113
Problems and questions
  answers, selected problems, 331–332
  capital cost, 5, 44–46, 60, 92–93, 100, 114, 134–135
  complete estimate, 248–250
  financial analysis, 226–227, 242–243
  operating cost, 148, 204–205
Profitability analysis, 213–227, 235–238
Published data, limitations of, 116–117

Raw materials
  captive, 159–160
  cost of, 158–160
Rentals, 182
Report preparation, 245–248
Retrofitting, 95–100
  accessibility, 96
  as-built drawings, 97
  control systems, 98
  electric power source, 97
  motor control centers, 97
  pipe-rack loading, 98
  safety considerations, 96–97
  scope of work, 99
Return on average investment, 220, 222
Return on investment, 219–222
Reugg, R. T., 242
Revenue requirement, 225–226
Royalties, 153, 182

Savings-to-investment ratio, 236–238
Scheduling, 110–112
Scope of work, 3, 47, 99

Semivariable costs, 150
Sensitivity analysis, 239–241
Seven-tenths rule, 9
Shipping costs, 203–204
Shutdown point, 156–158
Sitework, 75
Six-tenths rule, 9
Special construction, 85–86
Specialties, 81
Startup costs, 40, 207–209
Statement of work (*see* Scope of work)
Supervision costs, 178–179

Taxes, 56, 219
  real estate, 203
Terminology, cost engineering, 253–275
Thermal protection, 79
Timmerhaus, K. O., 12–13, 18

Uncertainty, 239–241
Utilities, 98–99
Utility costs, 161–164
Utility summary, 38–39, 162

Variable costs, 149
Venture worth analysis, 220, 224–225

Walker, C. G., 117–118
The Wall Street Journal, 159
Water costs, 163
Wessell, H. E., 165
Windows, 79–80
Wood, 78
Work breakdown structure, 62–63, 105–106
Workhours (*see* Labor)
Working capital, 40, 57